CAMBRIDGE LIBRARY

Books of enduring schol

Life Sciences

Until the nineteenth century, the various subjects now known as the life sciences were regarded either as arcane studies which had little impact on ordinary daily life, or as a genteel hobby for the leisured classes. The increasing academic rigour and systematisation brought to the study of botany, zoology and other disciplines, and their adoption in university curricula, are reflected in the books reissued in this series.

Memorials of Sir C.J.F. Bunbury

Sir Charles James Fox Bunbury (1809–86), the distinguished botanist and geologist, corresponded regularly with Lyell, Horner, Darwin and Hooker among others, and helped them in identifying botanical fossils. He was active in the scientific societies of his time, becoming a Fellow of the Royal Society in 1851. This nine-volume edition of his letters and diaries was published privately by his wife Frances Horner and her sister Katherine Lyell between 1890 and 1893. His copious journal and letters give an unparalleled view of the scientific and cultural society of Victorian England, and of the impact of Darwin's theories on his contemporaries. Volume 1 begins with a short autobiographical memoir written by Bunbury towards the end of his life, and ends with his marriage to Frances Horner in 1844. It demonstrates his interest in science, encouraged by his parents and continued at Cambridge and during travels in Europe and South America.

Cambridge University Press has long been a pioneer in the reissuing of out-of-print titles from its own backlist, producing digital reprints of books that are still sought after by scholars and students but could not be reprinted economically using traditional technology. The Cambridge Library Collection extends this activity to a wider range of books which are still of importance to researchers and professionals, either for the source material they contain, or as landmarks in the history of their academic discipline.

Drawing from the world-renowned collections in the Cambridge University Library, and guided by the advice of experts in each subject area, Cambridge University Press is using state-of-the-art scanning machines in its own Printing House to capture the content of each book selected for inclusion. The files are processed to give a consistently clear, crisp image, and the books finished to the high quality standard for which the Press is recognised around the world. The latest print-on-demand technology ensures that the books will remain available indefinitely, and that orders for single or multiple copies can quickly be supplied.

The Cambridge Library Collection will bring back to life books of enduring scholarly value (including out-of-copyright works originally issued by other publishers) across a wide range of disciplines in the humanities and social sciences and in science and technology.

Memorials of
Sir C.J.F. Bunbury

VOLUME 1: EARLY LIFE

EDITED BY
FRANCES HORNER BUNBURY
AND KATHARINE HORNER LYELL

CAMBRIDGE
UNIVERSITY PRESS

CAMBRIDGE UNIVERSITY PRESS

Cambridge, New York, Melbourne, Madrid, Cape Town,
Singapore, São Paolo, Delhi, Tokyo, Mexico City

Published in the United States of America by Cambridge University Press, New York

www.cambridge.org
Information on this title: www.cambridge.org/9781108041126

This edition first published 1891
This digitally printed version 2011

ISBN 978-1-108-04112-6 Paperback

MEMORIALS

OF

Sir C. J. F. Bunbury, Bart.

EDITED BY HIS WIFE.

EARLY LIFE.

MILDENHALL:

PRINTED BY S. R. SIMPSON, MILL STREET.

MDCCCXCI.

MEMORIALS OF SIR CHARLES J. F. BUNBURY.

CHARLES JAMES FOX BUNBURY was born on 1809
February 4th, 1809, at Messina, and was the eldest
living son of Col. Henry Edward Bunbury. Col.
Bunbury was serving at the head of the Quarter
Master General's department of the English Army
in Sicily. He had married in April, 1807, Louisa
Emilia, daughter of the Honourable General Fox,
who was in command of that army, brother of the
celebrated statesman, Charles James Fox.

Mrs. Bunbury writes in her journal: "On Saturday, February 4th, I had the delight of being the
mother of a very fine boy. I cannot express how
thankful I am for this blessing."

On Thursday, March 23rd, this child was
christened by the names of Charles James Fox,
which combined the names of his father's uncle,
Sir Thomas Charles Bunbury, and his mother's
uncle, Charles James Fox. Col. Bunbury finding
there was no chance of active operations of the
army at that time, obtained leave of absence, and on
August 15th embarked with his wife and infant
child for England, but it was not till towards the
end of September that they landed at Falmouth.

1809 He had not been a month in England before he
accepted the office of Under-Secretary of State for
the War Department.

I will now copy some extracts from Sir Charles'
own reminiscences, which he wrote for his wife
a short time before his death.

<hr>

AUTOBIOGRAPHY.

REMINISCENCES.

___

My earliest recollections seem to indicate a taste
which has in no-wise influenced my subsequent life.
In the earliest time that I can remember at all
distinctly, we lived in London, and my great delight
was seeing the Life Guards and Horse Guards pass
daily before the windows of my father's house.*
Indeed, my ideas were in those days very military;
and no wonder; for the time was near the end of
the Great War against France; my father was
Under-Secretary of State for War; my mother was
perfectly enthusiastic for the military and naval
services, and most of their friends and habitual
associates were either engaged in those services, or
officially connected with them. The conversation I
daily heard strongly impressed on my mind the
conviction, that there was nothing so great and

* I believe Sir Henry Bunbury then lived in Park Lane. F. J. B.

glorious as a brave and distinguished officer in the 1813 military or naval service of the Crown.

My mother took pains to impress on my mind the names of the heroes of the Peninsular War, and of the principal fields of battle. Wellington, Hill, Picton, Pakenham, Graham, Salamanca, Vittoria, and St. Sebastian,—were really familiar to my ears as household words.*

It was some time in the year 1813, that my father and mother gave up their residence in London, and we went to live at Mildenhall. My youngest brother Hanmer was born at Mildenhall, in December, 1813;† and a few days afterwards my father left us, having been commissioned by the Government to proceed to the Duke of Wellington's head-quarters, in order to confer with him on the condition and wants of the Army.

In a fragment which I have printed in the Memoir of my father, he describes the difficulty which he encountered, in reaching the port of embarkation in the midst of that uncommonly severe winter.

My own remembrance of this time is, as may be supposed, extremely indistinct, I had only a dim notion that I had another brother, that my father had gone away, and that my mother was unhappy.

---

* Sir Henry Bunbury in his latter years was talking to his doctor, Mr. Image, and asked him to guess what was the subject his son Charles was most strong in. Mr. Image guessed Botany, Geology, &c. Sir Henry said "Military Science." I have heard that General Sir Charles Napier said that there were few who knew so much of Military Science as Charles Bunbury.
F. J. B.

† His brother Edward Herbert was born in 1811, and his brother Henry William in 1812.

1815 I was a very delicate child, and therefore had little experience of the gay and buoyant spirits which make the chief happiness of childhood.

Mildenhall in those days was not exactly a cheerful abode, at least according to the impression which has remained on my mind. The house was much more closely surrounded, and shadowed by trees and bushes than it is now, for besides the great trees there was a thick sort of jungle shutting out the paddock, and the part where the greenhouse now is, from the house and lawn.

Within the house there was (as usual in old houses) a multiplicity of long, dark passages, of old melarcholy-looking staircases, of mysterious doors, and of huge closets, very well suited for hide-and-seek, but also very fit to suggest the idea of lurking places for *uncanny* beings. I have often thought, since I have read Byron's *Lara*, that the description of his hall has some resemblance to Mildenhall, at least as I then imagined it.

It has often struck me that the incidents of childhood, as one sees them dimly and imperfectly remembered, seen in fragments as it were, when one tries to recal them in later life, appear like isolated rocks, standing up here and there out of the sea which has devoured all that lay between them, or perhaps, like peaks rising abruptly out of a dense and wide-spread mist.

The great event of the year 1815, made a distinct and vivid impression, but an isolated impression on my mind. I well remember my father coming into my mother's dressing-room, and telling her the great 1817 news which had just come—how " Lord Wellington had fought a great battle and gained a great victory near Brussels, in which he had completely beaten Bonaparte and the French army "—but that one fact stands out like a conspicuous solitary peak, from the thick mist of oblivion in which all the rest of that year is hidden from me.

I have been told that from my birth until some time in the second year of my life, I was a healthy and promising child. The cause of the unfortunate change in my constitution I believe was never known, but it is certain that from the very earliest time that I can remember, I suffered grievously from a disorder, which not only caused me a great deal of physical pain, but affected my spirits, and had probably a permanent effect on my disposition. My mother, in the hope of finding a cure, consulted many doctors and tried many methods of treatment.

I have a dim remembrance of being carried to the Isle of Wight and other places on the South Coast; but at last, in the year 1816 or 1817 (I am not certain which), Southwold, on the Suffolk Coast, was fixed upon (by medical advice I suppose), for my sea-side residence.

Uninteresting as are the features of the place, the length of time I spent there sufficed to impress them strongly on my memory. I have always had a lively remembrance of the shingle beach, the bathing machines, drawn up in order under the low cliff, the open green between the town and the margin of the

1817 cliff, and the old cannon ranged on it.   This cannon
was given, it is said, by the Duke of Cumberland, of
Culloden fame.

Southwold has often re-appeared in my dreams,
even in recent years.

There then I spent two summers, partly with my
mother, partly under the care of a nursery maid.
I need hardly say that in the latter case, I was very
much bored ; but I was beginning to find a resource,
in some degree, against tedium, in a pursuit which
has ever since continued to render me the same
service.

I cannot very distinctly remember the beginning
of my taste for Botany. In its origin it was, of
course, merely the love of pretty flowers, which is
(I suppose) common to all unsophisticated children ;
and my mother, who was herself passionately fond
of flowers, encouraged this taste in me to the
utmost.

But she was not merely fond of pretty flowers ;
she had a considerable knowledge (much more than
a smattering) of botany, as it was then generally
studied in England, that is, according to the Linnean
system : she took great delight in communicating
her knowledge to me, and I had not less pleasure in
acquiring it.

LETTERS.

From Sir Henry Bunbury.

Mildenhall.
October 7th, 1817

My Dearest Boy,

I should not have delayed so long to thank you for the nice letter you wrote to me when I was at Wyddial,* but that since my return home there has been company in the house, and I have had no leisure. Your letter gave me pleasure, not only because I am always desirous to hear from you often when we are not together, but also because it gave me so satisfactory an account of the manner in which you employ your time and of the progress you make in acquiring useful knowledge. You are, as you always have been, a dear good boy, and I love you with all my heart. Write to me again, my dear Charles, and tell me what further you have learned or read, and tell me if you like and understand the little books I lent you, I mean the Scientific Dialogues. Your poor Mother has been in a sad anxiety for several days owing to one of your letters having miscarried, and if she had not been relieved this morning by the arrival of your little letter of Sunday, I verily believe she would have set off at full speed for Southwold to ascertain that you

---

* Wyddial was the home of Sir Henry's aunt, Mrs. Gwyn, (Goldsmith's " Jessamy Bride ") and her husband, General Gwyn. F. J. B.

1817 had not tumbled into the roaring sea which you describe so well.

We go to Bury tomorrow for several days, and I shall be very glad indeed when the Fair is over, and when I shall have returned again to our peaceful home. God bless you, my darling Boy.

I am, your affectionate Father,

H. E. BUNBURY.

P.S.—Edward can do the longest sum in Division without a mistake or difficulty.

AUTOBIOGRAPHY.

At Southwold I had the opportunity of seeing several wild plants, which I did not know before, particularly the showy yellow Sea-Poppy (*Glaucium luteum*) which is very abundant there, and the Sea Aster (*Aster Tripolium*) which decorates the salt marshes. My mother's botanical knowledge was not confined to British plants. Much of her early youth had been spent in the countries on the Mediterranean—at Gibraltar, in Minorca, and in Italy,* where she and her sister † had learned to know well the characteristic plants of those delightful regions.

---

* He meant *Sicily*. Those three places where her father, General the Hon. Henry Edward Fox, commanded. F. J. B.

† My Aunt was afterwards the Wife of Sir William Napier.

She used to show me the collections she had 1817 made of dried plants of those countries, for she dried them very well, only her specimens were too small and fragmentary. She took great delight in cultivating all the exotic plants she could in a small greenhouse at Mildenhall; and in an Alpine border in her garden there, she reared several scarce and interesting little plants, some of which I have never since seen alive; I remember particularly *Cornus Canadensis, Polygala, Chamæbuxus*, and some of the Alpine Primulæ.

By all these opportunities my love of botany was encouraged and heightened, till it became an actual passion.

My mother was acquainted with Lady Rous (afterwards Lady Stradbroke) and sometimes took me to stay for a little while at her house at Henham, within a few miles of Southwold. She (Lady Rous) showed a great deal of good nature and kindness to me, taking me with her through the gardens and hothouses at Henham, and lending me volumes of the *Botanical Magazine* and other fine botanical books to look through as I pleased.

My memory of my other studies (those which were to be considered as lessons) is less distinct; but I think it must have been in this year (1817) and the next, that I was taught French—taught to a certain extent by my mother, and afterwards, as to the strict rules and formulæ of grammar, by my father. To both I was indebted for instruction in the pronunciation of that language.

1818   Of public events of these years, one, and one only, is strongly impressed on my memory, this is the death of the Princess Charlotte. I very well remember that I was with my mother travelling post from Southwold to Mildenhall, and as we changed horses at —————— I forget what town, the master of the inn told my mother the sad news which had just arrived. She was shocked and affected by it in a way which (I well remember) surprised and puzzled me very much; and even after she had tried to explain to me the cause of her distress, I could but imperfectly comprehend it.

My malady was not cured by sea-air or by sea-bathing, therefore my father and mother, by medical advice, determined to try the effect of Malvern, which had begun to have a reputation in disorders such as that from which I suffered. This resolution proved very important to me, and the first visit to the watering-place with which I afterwards became so familiar, made a great impression. If I remember right, all my brothers were of the travelling party, which was numerous, for there were several servants.

My father drove his own horses, in an open carriage; and the journey from Mildenhall to Malvern took up several days. I think as many as five or six. I cannot recollect all the halts we made on the way; but I well remember that we saw Warwick Castle (where I was delighted with the armour, and with the supposed relics of Guy of Warwick) and the china manufactory at Worcester.

When we came near to Malvern, I was very much 1818 impressed by the appearance of the hills, which are indeed so bold in form, and rise so abruptly from the plain, that they may well be called minor mountains, the first of the kind that I can remember seeing. The season was spring, and I have a vivid recollection of the delight I felt in the luxuriant beauty of the country between Worcester and Malvern, the profuse blossoms of the orchards, and the abundance and variety of the wild flowers in the hedges. I had been used to little of this sort of beauty at Mildenhall, where there is a peculiar deficiency of *spring* flowers—and still less on the sea shore at Southwold.

One wild flower I well remember, which was new not only to me, but (strange to say) to my father and mother, and I was not a little pleased at being able to find out its name for myself, without help, in Withering's Botany. It was the yellow Dead Nettle, *Galeobdolon Luteum* of Withering, now called *Lamium Galeobdolon*.

I do not remember many other particulars of the year 1818. I remained at Malvern all through the summer, and most part of the autumn, under the care of my mother at first, and latterly of a maid servant.

Not long after they removed to Cheltenham, where in August, 1818, my sister Emily was born, the youngest of my father's children. This stay at Cheltenham was memorable for another reason. My father and mother took into their service a

1819 young woman of the name of Maria Fennell (as well
as I remember it) who had been cook in the house
where they lodged at Cheltenham ; they were so well
pleased with her, and found her so trustworthy, that
they left me under her care, not only when they
returned to Mildenhall in the autumn of that year,
but in each of the four years following.   She became
a permanent member of our family, and continued
in it, esteemed and respected by all, till December
in the year 1860, when she died in my house at
Barton, having survived my father scarcely more
than six months.

I well remember that it was during this visit to
Cheltenham that I saw for the first time the blue
Meadow Cranesbill, *Geranium pratense*, which is
plentiful in that neighbourhood, and was charmed
with its beauty.   I was taken back to Malvern year
after year, spending there the summer and autumn
until 1822, after which my health appeared to be so
well restored, that it was unnecessary to return
thither.   In June of 1819 my father and mother
took me with them in a delightful tour down the
Wye, and through a great part of South Wales.
We journeyed (as I have elsewhere written) in the
leisurely and comfortable fashion of those times,
when gentlemen really travelled, and allowed themselves time to look about them.

Ross, I remember, was the end of our first day's
journey ; and we spent the next two days on the
Wye, descending the beautiful river in a boat, and
landing wherever there was anything specially inter-

esting to be seen on the banks, particularly at 1819
Goodrich Castle, Coldwell Rocks, Tintern Abbey,
and Piercefield, sleeping the first night at Monmouth,
and the second at Chepstow. By Coldwell Rocks,
I mean the rocks overlooking the Wye, near the
village of English Bicknor, and close to the spots
marked in the Ordnance Map as Symond's Gate and
Coldwell. Our Guide called them the Coldwell
Rocks. It is certainly not without good reason that
Gray speaks with such enthusiasm of the Wye.*

At Malvern I had first become acquainted with
fine scenery, and under my mother's teaching had
learned in some degree (however imperfectly) to
appreciate and enjoy it.

On the Wye I was brought into some of the most
beautiful country in England, into the midst of some
of the most varied and delightful combinations of
river, rock, meadow, and wood ; and I hope that I was
not entirely insensible to their charms. But I learned
also how the enjoyment of scenery is enhanced by
the being able to associate it in one's mind with
great events, or interesting manners, or noble
characters. I was particularly pleased with the
sight of Goodrich and Chepstow Castles, and
Tintern Abbey ; for I had before this read with my
mother some of Walter Scott's poems, and some
romances of chivalry, and I was delighted to see in
actual substantial reality some such buildings as
those of which I had read.

---

* *See* Gray's letters, *ed.* Mason, letter of May 24th, 1771 (to Dr. Wharton.)
F. J. B.

1819   From Chepstow we proceeded to Newport on the Usk, Cardiff (from whence we took a long and fatiguing drive to see Caerphilly Castle), so on to Neath, Swansea, up the beautiful vale of the Towy to Llandovery, to Brecknock, Hereford, and back to Malvern by Ledbury; the whole tour having been rich for me in enjoyment and instruction. The Castle of Caerphilly, I thought sublime.

Through the whole tour my father and mother were constantly watchful both to promote my enjoyments, and to excite and guide my powers of observation. We found in the same town too, uncommon plants, which I have hardly since seen alive—*Campanula patula* near Usk, and *Linaria repens*, somewhere on the borders of Caermarthen and Brecknockshire.

The second tour was made in the year 1820, when I was eleven years old, and, with my father and mother visited some of the most interesting places in Derbyshire. What I best remember of this, was our stay of several days (a week or more I think) with old Mr. and Mrs. Morewood, at Alfreton Hall. The good old couple were excessively—I may with truth say *excessively*—kind to me.

I never saw Alfreton again till you and I visited it together in 1877.

After leaving Alfreton, we spent some days very pleasantly at Matlock, where I admired the rocks, and was very much interested in the caverns and the mineral shops.

My father had now become a zealous student of

mineralogy and geology; my mother shared in his 1820 taste, though with less knowledge; and I acquired some liking for their pursuit, although my especial predilection still was for botany. At this time, however, I learned from my father to understand something about strata, and mineral veins, and the differences between limestone and "*toadstone*" (amygdaloid), and the controversy between the Wernerian and Huttonian schools of geology, in which he took an eager interest, having become a zealous Huttonian.

After Matlock, I think the next place where we made a halt was Bakewell, for the purpose of seeing Haddon Hall. I was a good deal impressed by the antiquity and picturesqueness of this building, but I was much more struck by the caverns of Castleton, where our Derbyshire tour came to an end. Indeed, I still think that the great Peak Cavern at Castleton, with its grand entrance, like a vast arched gateway, in the face of a lofty and naked cliff, is the finest specimen of a cavern that I have ever seen, though I am told that the cave of Adelsburg in Austria is much finer. Homboldt, who, when treating of caverns in his *Relation Historique* especially mentions this of the Derbyshire Peak, seems to have been greatly impressed by its singularity and grandeur.

My chief botanical acquisitions in Derbyshire, in the tour of 1820, were the pretty yellow mountain pansy, *Viola lutea*, very abundant about Matlock and Castleton; that curious, pale, fleshy root-parasite the *Lathræa Squamaria*, which we found in the

1820 woods, below the High Tor at Matlock, and which
I have never since seen alive; and the *Arenaria
verna*, and *Thlaspi Alpestre*, remarkable for flourishing especially amidst the mineral heaps thrown out
from the lead mines.

At Malvern I learnt by degrees more and more to
enjoy the beauty of scenery. I was very fond of
standing on the ridge of the hill, which commanded
glorious views in both directions:—on one side over
the fertile and highly-cultivated plain of Worcester-
shire, divided so regularly by hedges into square
fields that it looked like patchwork, with the bold
mass of Bredon Hill standing up like an island in
the middle of it, and the Cotswolds bounding it in
the distance;—on the other side, over the much
more varied surface of Herefordshire, with its rich
woods and range beyond range of hills, in picturesque
gradations of distance to the far mountains of
South Wales.

Another pleasure I well remember in those
Malvern days was to look down from the ridge of
the hills, on a fine autumn day, and see huntsmen
and hounds pursuing their sport amidst the copses
and fields below, and listen to the cry of the pack as
it came up mellowed by distance.

I do not well remember in what year it was, but it
was early in my Malvern life—when I was nine or
ten years old—that my father began to read aloud
to me select bits of Shakespeare, an author with
whom I then first made acquaintance; and finding
how much I enjoyed it, he and my mother per-

severed in the practice, till I became familiar with 1820
many of the best plays. Of course, at that age I
was not able really to appreciate the glorious
poetry ; what I relished, as I remember, was first,
the incomparable fun and drollery of Falstaff, and
the merry contests between him and Prince Hal;
next the splendid chivalrous character of Hotspur,
and a little later (when I began to be pretty well
read in Plutarch) the grandeur of Coriolanus and
Cæsar, and Brutus and Cassius. Afterwards, when
I was left alone with my nurse Maria for a winter
(that of 1820-21 I think), Shakespeare was one of the
books which my mother left with me for my amusement and instruction ; and I did study it diligently.

Walter Scott's poems were also among my most
favourite readings in those days. My mother
delighted in them, and would read or recite them
to me over and over again, dwelling on all their
merits and explaining any difficulties, with such a
loving care that these poems have ever since been
associated in my mind with the idea of her. They
were the first poetry (with the exception perhaps, of
Pope's Homer) that I ever really enjoyed, and it
may be owing partly to those early predilections,
that I even now read Scott more often, and relish
him more, than people in general seem to do.
With his novels I made acquaintance rather later,
though still in my Malvern days ;—" Ivanhoe " I
remember, was the first I read, " Waverley " the
second—both with my mother—and both I rather
think in 1820.

B

1820   I delighted in them, and I still vividly remember the enthusiasm with which my mother read Flora MacIvor's Appeal to the Clans and what gratifying approbation she bestowed on my attempts to read like her.

Langhorne's Plutarch was another of the books with which I became well acquainted during my Malvern days, and it interested me exceedingly.

Plutarch's " Lives " is a book I should certainly commend all boys to read, as I think it eminently fitted to cultivate noble and exalted sentiments.

My dear mother, who was deeply, earnestly, unaffectedly religious, not only attended closely to this branch of my education when she was with me, but impressed strongly upon me the duty of carrying on in her absence, the religious observances of the day.

When I was at home at Mildenhall (for I was recalled thither for a part of every one of these years) my studies, principally in French and arithmetic, were pursued more steadily and systematically, under my father's direction, and I believe I made tolerable progress in them. Arithmetic I detested most cordially, and have continued to do so ; to French I had a much less decided dislike, even to the grammar. Of Latin I learned nothing till I was either twelve or thirteen years of age—I am not sure which.

Edward and I pursued our studies—*pari passu*— side by side, for though he was two years and a half younger than I, his studies had been free from

those many and long interruptions which ill-health 1820
and my Malvern banishments had occasioned in
mine, and therefore he was at least as far advanced
in learning as I was.

It was in December, 1819, that my little sister
Emily died, after a very short illness at Mildenhall.

I did not understand till afterwards, the profound
grief which this death occasioned to my parents,
and especially to my mother.

Besides the readings which I have already
mentioned, one of my principal amusements, and
indeed occupations, during this time of my life, was
drawing. This I had begun very early, indeed, my
memory hardly goes back to a time when I did not
try to draw animals or flowers, and in particular
to copy the prints of such objects in books. I well
remember how once, at Mildenhall, I was discovered
embellishing the white skirting-board of the nursery
with drawings of birds and beasts; a proceeding for
which I was scolded by the nursery-maid, but
applauded and encouraged by my mother, when the
case was referred to her. At a somewhat later time,
I was indefatigable (as, indeed, was Edward also)
in copying birds from Bewick and the copy of that
delightful book which was in my father's library,
and is now in mine, bears marks (I fear) of careless
usage on our part. In the lonely winter at Malvern
I had no such resources, but I had the courage to
attempt to draw birds from nature; and our land-
lady good-naturedly got her husband to shoot birds
for me to draw. I remember drawing a Ring-Ousel,

1820 and a Brown Owl, which were thus procured for me. I had in early days become fond of romances of chivalry, and delighted in reading (in the English translations, of course) " Amadis de Gaul," and " Palmerin of England ", as well as the romances included in Percy's and Ellis's collections.

I do not remember, in what year it was, but probably either in 1819 or 1820, that I met with a translation of the famous " Battle of the Frogs and Mice " (formerly ascribed to Homer; but, of course, really belonging to a much later time). The idea took my fancy exceedingly, and after reading the story over till I was quite familiar with it, I set to work to make drawings of the battles such as I imagined them from the poem, arming the warriors as they were there described.

Williamson's " Oriental Field Sports," and Daniell's " Rural Sports," suggested the subjects of very many of my drawings, the former of these books I had seen at Alfreton, the other I was allowed to look over again and again, in the house of a kind friend (Lady Wigram) at Malvern ; and thus it came that my Mousicarians were sometimes engaged in shooting grouse or pheasants, sometimes in spearing wild boars or tigers.

It is rather amusing to me to recollect that my imaginary people were thus represented as entirely devoted to two pursuits (war and the chase) which have had in practice no attractions for me.

Botany continued to be my favourite study, in fact my *hobby*, throughout the early years of my life at Malvern, though towards 1821, a taste for 1821 ornithology began to assert some rival claims. But Botany had the great advantage of being much more easily pursued in a practical way.

Of society, during the time I had to spend at Malvern, I can say little.

I have a dim and hazy sort of memory of old Lady Harcourt and of a little Miss Bloomfield. Miss Bloomfield was about my age, I heard some years afterwards that she died before she was grown up.

Lady Wigram was very kind to me.

I remember Mr. (afterwards Sir Charles) and Lady Charlotte Lemon; though indeed, my recollections of that date are more of Lady Charlotte than of her husband. Many years afterwards, I learned to know and to love him, by my delightfnl experience of days spent in his house at Carclew, in 1841; one of the most agreeable reminiscences of my bachelor life. He had then been many years a widower, and childless; and in the church there I saw tablets to the memory of Lady Charlotte and their two children.

I think it was in the year 1822—the last summer which I spent at Malvern—that Mrs. Lock and her two daughters came thither, and, of course, we (for my mother was there, and I think Edward also) saw a great deal of them. The elder daughter, Cecilia, was very handsome, and I well remember that she was the first lady whom I ever admired for her beauty, and for some time she figured in my imagina-

1822 tion as the representative of Oriana and other heroines of romance. Her sister Lucy (Mrs. Ellice) was not beautiful, but, as you well know, very clever and amusing; and both sisters were very good-natured to Edward and me, and we liked them very much.

I knew French tolerably well at ten years old; but I think it must have been in the winter of 1821-22 that my father began to give me lessons in Latin; and that these were resumed after I returned home from Malvern in the latter part of the summer of 1822; while I was at Malvern in this last year, a clergyman in the neighbourhood, a Mr. Deane, was engaged to read Latin with me for an hour or two in the day.

At last, in 1822, it was decided that my health was sufficiently established for Malvern to be no longer necessary, and, accordingly, in the autumn of that year, I took my leave of the place where I had spent so considerable a part of my childhood. From that time, my old complaint almost ceased to trouble me; and I have since (especially in my old age) had reason to be thankful for generally good (though not superfluous) health.

In the autumn of that same year, 1822, Mr. Matthews (Frederick Hoskyns Matthews) whom my father had engaged to be private tutor to me, and my brothers, arrived at Mildenhall. He remained with us in that capacity for five years, till we went abroad in the autumn of 1827; and his character and instructions have remained deeply impressed on my mind, so that I may fairly consider him as one 1822 of those who have had an influence on my life.

He was one of the youngest (I rather think actually the youngest) of a very numerous family in Herefordshire; his father was a partner in one of the principal banks at Hereford: his eldest brother, Charles Skinner Matthews, was one of Lord Byron's intimate associates at Cambridge, and was looked upon as a young man of high promise, though of considerable eccentricity; but he was drowned in the Cam before he had done anything to justify the hopes of his friends. Another brother, Henry, was the author of a pleasing, though not very profound work, the "Diary of an Invalid." Our Mr. Matthews was a little more than ten years my senior; he had been educated at Shrewsbury school under Dr. Butler, and at Trinity College, Cambridge; how it was that he failed to obtain a Fellowship in the latter, I do not know.* He knew thoroughly well what he professed to teach us— Latin and Greek and Mathematics—and I think he was a very good teacher. In classics at least, I know that his instructions have remained very much impressed on my mind; so that particular passages of the ancient authors frequently recal to mind his particular phrases and modes of construing.

Mr. Matthews had been a contemporary at Trinity College of Macaulay and Praed, and acquainted (though I think not intimate) with both

---

* I believe that he had a decided and conscientious dislike to "taking Orders" in the Church.

1822 of them, and he talked much of the brilliant talents and grand promise of both. But it seems rather strange to us now, looking back on their actual performances, to know that they were then considered as rivals.

Mr. Matthews was a good, quaint, simple-hearted man, a little hot and hasty in temper perhaps, but then, I daresay we often gave him a good deal of provocation ; but he was very sensible to kindness, and always expressed himself as very grateful to my father for the way in which he had been treated at Mildenhall and Barton. Mr. Matthews was well read in English literature, at least in that of the eighteenth century, to which he was very partial ; indeed, his taste was altogether regulated by the canons of criticism which prevailed in the time of our grandfathers ; he was steadily devoted to the "classics" in English as well as in Greek and Latin. Of French, I think he had but an imperfect knowledge, and of other modern languages none at all.

He, after we went abroad in 1827, became tutor in the family of Mr. Marmaduke Wyvill of Constable Burton, in Yorkshire. How long he remained there I do not know, nor how long afterwards it was, that one of his elder brothers took him into partnership in a bank in the city of Hereford, which I have already mentioned.

After we had parted in 1827, I never saw him again till we (you and I) visited him at Hereford in 1855, when he greeted me very cordially. He had

then become (if this expression be not too inaccu- 1822
rate) sole partner in a seemingly thriving bank, and
was comparatively a rich man ; but he received me
in a very modest lodging, and pointed out to me
with pleasure, and pride, how like a room in college
he had made it. When some years afterwards the
concern of the bank turned out unfortunately, and
he was reduced to comparative poverty, I heard
from some who saw him often, that he did not
merely bear this reverse of fortune with equanimity,
but was actually pleased with it, enjoying his
release from all the cares and trammels of business
and wealth. Under Mr. Matthews' tuition then, we
settled into a regular, steady, continuous course
of study, in which Edward and I soon made
satisfactory progress. I cannot honestly say that,
at first, the new system was very agreeable to us,
or that we and the tutor got on very well together,
without at least, occasional difficulties. By degrees
we got used to one another's ways ; he was pleased
with Edward's and my diligence and progress ; and
at last we became very good friends. I had always
been of sedentary, and in some degree of studious
habits; and though the subjects of my voluntary
studies were very different from those to which my
attention was restricted by our tutor, yet my instinct
of docility, and my strong desire to please my father
and mother, were sufficient, I believe, to make me
a satisfactory pupil. Edward, unlike me, had the
physical strength and energy, which might have led
him rather to cultivate the favourite sports of boys ;

1822 but his passionate love of knowledge made him
devote himself to books with as much ardour as if
he had cared for nothing else. Under Mr.
Matthews' tuition we two advanced with a pretty
equal pace ; we soon passed from Latin into Greek,
and in the course of the five years we had gone
through many of the principal works in both
languages. I can by no means give a complete list
of the books we read ; but I well remember the first
six books of the Æneid, nearly the whole of
Horace's Odes, and most of his Satires and
Epistles ; Ovid's Epistles and many portions of the
Metamorphoses ; the tenth and some other Satires
of Juvenal ; Cicero on "Old Age ; " (I do not
remember that we read with Mr. Matthews anything else of Cicero than the "De Senectute ;")
many books (I am not sure how many) of the Iliad ;
the "Prometheus" of Æschylus ; the "Œdipus
Tyrannus," "Ajax," and some other plays of
Sophocles ; several of those of Euripides ; select
portions of Theocritus, Bion, Moschus, Anacrean.

Mr. Matthews, as a true orthodox scholar of the
public school type, of course thought it necessary
that his pupils should go through the drudgery of
making Latin verses—*invita Minerva.* I hardly
know that any part of the time of my youth was
more thoroughly wasted than the hours I was
compelled to spend on these Latin verses. I had
no particular talent or inclination for mathematics ;
but neither had I any special incapacity for it ;
indeed, I liked algebra well enough ; and giving my

best attention to Mr. Matthews' instructions, before he 1823 left us, I had entered on the mysteries of the Differential Calculus. In fact both Edward and I had studied so well that I think I may say, if we had gone at once from Mr. Matthews to Trinity College, we should have been qualified to take very good places at Cambridge.

Of course, these studies had not so entirely occupied my time or thoughts as to exclude all attention to my old favourite pursuits of natural history.

Through the greater part (if not the whole), of 1822, and perhaps a part of 1823, Entomology continued to be the favourite; but before the summer of 1823, I had completely dropped the study of insects, and given myself up to that of minerals, in which my father could heartily sympathise with me and assist me, as it had for several years been one of his hobbies. I soon grew quite *fanatico* for this pursuit—as much absorbed in it as I have ever been in Botany. As much of every day as could be spared from "lessons," meals, and necessary exercise, I devoted to my dearly beloved minerals, studying them not only in books (Aiken, Phillips, Jamieson), but still more in my father's cabinet; I must own that my mineral studies were not very profound. I was not mathematician enough to follow Haüy's researches into the formation of crystals, and though I dabbled in Chemistry so far as to examine many fragments of minerals with the blowpipe, I did not go deep into either the theory or the practice of that grand science.

1824   My father at the same time took a wider range and studied Geology with as much ardour as Mineralogy; but I did not follow his example till a later time. In the spring, or early summer of 1823, we boys had the great pleasure of accompanying our father and mother in a tour to Hampshire, the Isle of Wight and Weymouth.

Sir Thomas Charles Bunbury had died in March, 1821, and my father had succeeded to the estate; but he found everything at Barton in such a neglected state, and the house itself requiring such extensive repairs and improvements, that it was impossible to make it our home till after a considerable time; and in fact Mildenhall continued to be our residence until the summer or autumn of 1824.

The autumns and winters of these years were gay times; my father and mother then received a great deal of company, principally their friends from London, with whom their acquaintance had begun in their Mediterranean days. Among the guests there were not a few who had been prominently engaged in the Great War against France. But I can remember only names, and but few even of them; principally Sir Edward Codrington, Sir Henry Torrens, Sir Herbert Tayler.

My father's old aunt, Mrs. Gwyn, was of course a regular visitor at our house at Mildenhall.

I need not tell you about Mrs. Bucke and her daughter Augusta, whom both you and I knew so well when *we* were living at Mildenhall. Laura Bucke (Mrs. Adams) was very beautiful, but yet I

remember to have heard that my uncle, Sir William 1824
Napier (and some other gentlemen) admired the
mother even more than the daughter.

The impression which has remained on my mind of
those years—from the autumn of 1822 to that of
1827, or at least to the beginning of my mother's
ilness, is that of a happy, peaceful time, with no
material vicissitudes, no remarkable events, nothing
(as far as I know) to disturb the tranquil current of
our days. My health, though not robust, was good ;
I made quiet and satisfactory progress under Mr.
Matthews' teaching, in those studies which my father
wished particularly me to pursue ; I had plenty of
leisure and opportunity for my favourite pursuits ;
my intellect was gradually expanding ; I had no cares
or anxieties ; I was as yet a total stranger to the
plagues of money ; I met with uniform kindness from
both my parents, and with more than kindness from
my mother. Indeed, I hardly know how I could
have been otherwise than happy with such a mother.

I can never regret that I have to look back on my
boyhood and early youth as a happy, happy time,
exempt from the severities of old-fashioned discipline
as well as from the anxieties of mature life.

To the best of my knowledge, my mother never
spoiled me or my brothers ; she never allowed us to
do what she knew to be hurtful ; she never tolerated
any confusion between right and wrong ; never showed
passion or selfishness in reproving us for our faults ;
she did not confuse and worry us by a multiplicity
of petty precepts and prohibitions but when she

1824 once laid down a rule on some serious and important matter, she required it to be steadily observed.

In the latter years of our life at Mildenhall, 1822-24, I was a very keen sportsman in an innocent way; not that I was yet allowed to carry a gun, but my father who was in those days fond of shooting, and a very good shot, very often allowed me to follow him when he went out with his gun and dogs; and I took great delight in these rambles.

Battues were hardly yet in vogue in those days, at least my father did not care for them; what he enjoyed was partridge shooting, and a good deal of the pleasure, I think, consisted in watching the skill and sagacity of the dogs. I have already mentioned my assiduous study of Daniell's "Rural Sports" and similar works; and what with such books, and with the attending my father over the fields, I became in imagination a keen sportsman, and longed for the time when I should be allowed myself to carry a gun and to kill game. That permission was to be granted when I should be sixteen years old. My sixteenth birthday came, and my passion for field sports had quite evaporated!—and it never returned.

In the spring of 1824 (I think it was), my father and mother took me and Edward and Henry with them to London, where we remained, I think, two or three months, in order that we might have lessons in dancing and riding, and also that we might be introduced to some of their friends.

What I did enjoy, in this visit to London, were

repeated visits with my father and mother to the 1824
British Museum, where I passed hours of indescribable delight (and I could gladly have spent days) in
studying the minerals. There were, indeed, even
then, a magnificent collection, and well arranged,
and my father, having made acquaintance with Mr.
Children, who had then the charge of that department of the Museum, enjoyed the advantage (which
I was allowed to share) of studying them in the
tranquillity of " private days."

Among my father's and mother's friends whom I
saw during this visit to London, Dr. and Mrs.
Somerville were those whom I remember as most
particularly kind and friendly to me, and whom I
liked the most. I was not at that time qualified to
understand and appreciate, the extraordinary intellectual powers of Mrs. Somerville, but I could feel
the peculiar gentleness and simple kindness of her
manners, and I learned much from her and Dr.
Somerville in my favourite study of Mineralogy, to
which, in those days, they also were much devoted.
They made me very happy by their great good-
nature in giving me a large number of specimens of
minerals, several of them rare and valuable; these
still remain in the cabinets at Barton. One day, I
remember, my father took me to the rooms of the
Geological Society, and introduced me to Dr.
Buckland and Mr. Warburton, with whose names I
was already well acquainted; indeed, I had read
and admired Buckland's " Reliquiæ Diluvianæ "—
and was very glad to see such a famous man.

1824   In the summer or autumn (I am not sure which)
of 1824, we removed to Barton, and settled our-
selves in our new home.

I well remember that when I was there in my
childhood—in Sir Charles' time, there was no
pleasure ground adjoining to the house, but the open
pasture of the park reached up to the walls, and
there was only a line of posts and chains to keep off
the horses and cows from the windows. The
present pleasure ground had been enclosed and
planted by my father, before we settled at Barton,
and made a great difference in the appearance of
the place, though of course the shrubs were very
low in comparison with what they now are.

Not long after our arrival at Barton, Hanmer one
day took it into his head to *fish* in the trumpery
little pond which had been formed in a corner of the
new pleasure ground; and he hooked an unfortunate
newt. We therefore looked into Pennant's British
Zoology (our only authority in those days on
zoological matters) and we found that, according to
that authority, the creature was a kind of *lizard*.
We therefore settled forthwith that the pond should
be called *Lizard Lake*, and *Lizard Lake* it has been
ever since.

My mother found at Barton a garden ready
walled, quite sufficiently large both for flowers and
kitchen garden, and well adapted for those purposes;
so she soon formed a very pretty and well-arranged
flower garden, in which she cultivated many
beautiful and interesting things. She managed to

grow successfully many fine tropical plants ; a great 1824 part of them raised from seeds sent to her by friends in distant countries. Indeed, one plant, the *Justicia speciosa Roxb*, (*Peristropha Nees*) was recorded by Sir William Hooker in the Botanical Magazine, as having made its first appearance in England in my mother's hothouse. This I remember was sent to her from St. Helena ; having no doubt been raised there in a garden, for its native country is Bengal. Other interesting plants which flowered in the hothouse at Barton were—*Cæsalpinia* (*Poinciana*) *pulcherrima* (which deserved its name), *Calathea Zebrina*.

My mother took in Loddiges' " Botanical Cabinet," a work which was of great use to her, in suggesting plants worth ordering for her garden, besides supplying concise instructions for their culture. Of strictly scientific botanical information it contains none, but the plates are (in general) remarkably good, spirited characteristic likenesses of plants, though usually on a small (sometimes a very small) scale.

My father had begun the Arboretum at Barton almost immediately after he had come into possession of the place, forming it out of three paddocks ; and many of the finest trees in it, were planted between 1822 and 1827. What is called the American Garden (which is in effect a part of the arboretum) got that name because it was intended more especially for the cultivation of what were then known to gardeners as American* shrubs, including

* Of which, by the way, very few are natives of America.

1824 Rhododendrons, Azaleas, Kalmias, Magnolias, and many others, for the most part of the mountains and forests.

My mother took a great interest in this part of the garden, and selected many of the objects intended for it ; but the success was very imperfect, it was not understood how much the growth and health of those shrubs of the Heath family is dependent upon a particular soil, and that chalk (of which there is a great deal in the soil of Barton) is unfavourable to them. Some plants of *Rhododendron Ponticum*, for which my father prepared places by removing the soil of chalky gravel to a considerable depth, and filling up the hollow with soil and peaty earth brought from Cavenham Heath, have thriven very well, and flower abundantly every year, but the others, even of that species, are very poor, and the Kalmias have barely kept alive, without once blossoming.

The *Magnolia acuminata* has grown into a beauti ful and stately tree, while its congeners, *Magnolia tripetala*, and *Magnolia auriculata*, are far from thriving.

The *Carolina Allspice, Calycanthus floridus*, which was one of the shrubs planted at that time, has thriven very well.

During these years at Barton, until my mother fell into bad health, we saw much company—not only (as at Mildenhall) old friends from London, but also numerous families of the neighbourhood. Of course our studies under Mr. Matthews were not interrupted on this account; the hours devoted 1824 to classics and mathematics were as regular and as strictly observed as before.

The Blakes, being cousins of my father, were the first of the country neighbours with whom we became intimate, especially after William had been presented to the living of Barton.

Other neighbours of whom we saw much were Mr. and Mrs. John Powell—*he* was a quiet, sensible, painstaking magistrate, and my father was somewhat drawn towards him by agreement of political opinions, for Mr. Powell was a Whig—a rare animal in those days in Suffolk.

Mrs. Powell was very pleasing, gentle, and ladylike, and my mother liked her very much.

But the most interesting by far, of our neighbours in those days, and the one whom my father and mother valued the most, was old Mr. Hasted. I do not know that, at the age I then was, I was qualified to appreciate the real beauty of his character; but I liked his kind and gentle manner, his varied knowledge, and his readiness to impart it; I liked to visit him, and to see his collections of minerals and plants, which he was always willing to open to me.

I return to the early days of our life at Barton— the autumn and winter of 1824 and spring of 1825. My old love for botany had returned in full force, and I began that autumn and spring to search eagerly for plants in my walks around Barton. In the winter and spring of 1824-25 I first took up a new branch of the same science—the study

1825 of Mosses, which I soon began to find very attractive.

The neighbourhood of Barton is by no means rich in mosses; being nearly flat, and everywhere highly cultivated, without rocks or bogs, or old and extensive woods. But, thanks to our moist climate, a good many of these beautiful little plants are to be found almost everywhere in England, and I was tolerably diligent in searching for such as were within my reach. My dear mother had made me a present of a small but very useful microscope, and after copying a great number of the plates in " English Botany " I began to draw some of the mosses from Nature; I succeeded well enough to be encouraged to persevere, and went on till my collection of drawings became considerable. I cannot describe the pleasure I took in these pursuits; even at the present day, the sight of my old specimens, and my old drawings recalls something of the old enjoyment, which I experienced in those happy times.

In the spring of 1825, my anxious mother's fears were awakened by some symptoms which were thought to threaten a return of my old complaint. She determined to try the efficacy of sea air and sea bathing; Cromer was the place fixed upon, and she took me thither in May or June, 1825; my brother Henry being also of the party as well as our cousin Bessie,* whom she at that time had taken under her care.

* Afterwards Countess of Arran.

My studies were not entirely interrupted during 1825 this interval of absence from home; we took with us a Lexicon and some other books, and I remember I read at Cromer part of the "Ajax" of Sophocles, and corresponded with Edward on our respective readings in Greek.

I found near Cromer the *Bartarmia pomiformis*, but no other moss which was new to me.

In July my father joined us there, and we returned home with him, by a very circuitous route, spending some days at Yarmouth, and one or two at Aldborough. These were to me delightful days, above all the one at Yarmouth, when my father took me with him to make the acquaintance of Dawson Turner, whose reputation as a botanist was so familiar to me, that I looked on him with absolute reverence. A man who possessed such a knowledge of botany, and such a beautiful botanical library, appeared to me, one of the greatest and most fortunate of men. He was extremely kind to me, and made me a really valuable present of his volume on Irish Mosses—a small book indeed, but quite a gem in its way.

Another day of extreme enjoyment to me, was when my father took us out to some miles from Yarmouth, and leaving the carriage in the road we we spent some hours in botanizing in the Fen, near Fritton. It was the first opportunity I had had of botanizing in a genuine undrained fen, and it was one remarkably rich in interesting plants. I was indeed delighted with many which were novelties to

1826 me at the time, but which I afterwards found to be
abundant near Mildenhall, such as *Comarum palustré*,
*Ranunculus Lingua, Myriophyllum verticillatum, Utri-
cularia vulgaris*. But there was one curious plant,
which though it grew in profusion in that fen, I have
never seen since—the *Stratiotes aloides*.

Osmunda Regalis was another plant which I saw in
these Fritton bogs for the first time, and which I
never could find in West Suffolk. It is a rare plant,
I believe, in the Eastern counties generally, but
grew in abundance, and very large and luxuriant, in
the bogs of which I am now speaking.

We went to Malvern late in the autumn of 1825;
my mother and I together with my brother Henry,
and cousin Bessie.

In the spring of 1826, my father and mother took
me to " Woods," at that time a celebrated shop for
books on Natural History ; and they made me a
present of a really valuable book, Dillenius' celebrated work on Mosses. I need not say that I
studied it assiduously. In the summer of the same
year, Edward and I enjoyed a delightful tour with
our father and mother. We travelled, as before,
leisurely, making short day's journeys, and
stopping by the way wherever there was anything
interesting to be seen. Our first day's journey
(July 30th) reached no further than Cobham ; the
reason of our stopping there, was, that my
father and mother wished to see their old friends,
Sir Graham and Lady Moore, who were living
there.

The 31st we went on by Guildford to Farnham ;— 1826 the 1st of August to Winchester.

The 2nd of August we went on by Romsey, to Salisbury, where we spent the 3rd. The 5th of August we made a long day's journey, by Yeovil, Crewkerne, Chard, and Honiton, to Exeter, and this day I met with the first botanical treat that I had had on this journey. Just as we were about to begin the descent of Yarcombe Hill, between Chard and Honiton, my father and mother observed a bit of swampy ground, not far from the road, which looked botanically tempting ; we all left the carriage, and went to explore it, and in a few minutes I found three plants which were quite new to me.

Narthecuim ossifragum (with which I was especially delighted), *Drosera longifolia*, and *Rhynchospora alba*.

The next day we went from Exeter to Dawlish, and at the latter place we spent the time delightfully till the 17th of August, being the guests of Colonel (Sir George) Napier and his sister Emily ;* my real acquaintance with them began at this time ; both of them were as kind as possible to me. With Aunt Emily I had a special bond of agreement, as she had a real love of Botany.

I struck up a great friendship with the two girls, Sarah and Cecilia—a friendship never to end but with life—if then. Both were beautiful in their different ways. Sarah, who was then twelve years old, had even then the gentle, pensive, peculiarly

---

* Afterwards Sir Henry Bunbury's second wife.

1826 fascinating style of beauty which was so remarkable in her later years, and which so well corresponded with her lovely mind. Cecilia (aged seven) was the merriest little creature imaginable, full of spirits and fun.

Sarah had a passionate love of music, and a remarkable talent for it ; Cissy at that time did not care for it, and when the grand performances began, she used slyly to entice me away to play in the garden. My father seems, judging from his journal, to have been rather disappointed in the coast scenery about Dawlish, but he found some objects of geological interest in the cliffs of red sandstone and conglomerate (Triassic) and in the overlying Greensand of Haldon. I spent a great deal of time very much to my satisfaction in botanizing, finding many plants which were new and interesting to me. *Rubia peregrina,* in the hedges ; *Senebiera didyma* on the low ground near the village ; *Euphorbia Portlandica,* on the Warren ; *Lycopodium clavatum** on Haldon ; *Scutellaria minor,* in boggy spots on Haldon. Our delightful stay at Dawlish came to an end on the 17th of August, and we proceeded that day to Totness on the Dart, making a little deviation from the direct road, to see Berry Pomeroy Castle.

The 18th we went by Ashburton to Ivy Bridge, and on the 19th to Plymouth, seeing much of the fine and interesting country about the skirts of Dartmoor.

---

*This certainly is common enough, but it so happened that I had never before seen it growing.

We all were delighted with the scenery of Ivy 1826 Bridge, and I was not less so with the profusion and variety of Mosses growing on the rocks beside the stream. Between Ivy Bridge and Plymouth we went some miles out of the direct road to visit my father's old friend Sir John Colborne (afterwards Lord Seaton) who was living at a place called Puslinch. We stayed at Plymouth the whole of the 20th.

August the 21st we went by Tavistock to Launceston, where we were much struck with the singular character of the Church—built entirely of granite, and with every single stone, elaborately and variously sculptured on the outside. The old Castle was also observed. My father had intended to make an excursion from hence to Tintagel, but was discouraged by reports of the badness of the roads; for which reason we went back to Oakhampton, and there we remained till the 27th, finding it a convenient starting point for several excursions into the recesses of Dartmoor. We all found much to interest us in that wild country.

I found a great deal to interest me and my mother in the botany of the country around Oakhampton, especially in the wooded dells and along the banks, whether rocky or boggy, of the streams. Above all I was delighted with that exquisite little plant the *Campanula hederacea*. I admired also the *Osmunda regalis*, which I saw in great abundance, and in various stages of growth; I had here the opportunity of observing young seedling plants of the

1826 *Osmunda*, which look very different from the mature
fern, so that some of the old botanists took them for
altogether different species.

I found besides *Lycopodium Selago, Lastrea, Oreop-
teris, Pinguicula lusitanica*, and many Mosses.

From Oakhampton we went on the 24th of August
to Clovelly, and at that most beautiful and singular
place we spent the 29th. We were all very much
impressed by its beauty and peculiarity.

Our next halt was at Ilfracombe, which we reached
(by Bideford and Barnstaple) on the 31st of August;
and at Ilfracombe we stayed three days, so much
were my father and mother delighted with the
scenery. Our Western tour, which I enjoyed so
much that I still remember it with great delight,
terminated at Bath, from whence I suppose my
father and Edward proceeded directly homewards,
while (I know) I went with my mother to spend a
few days with my Uncle William, and Aunt Caroline,
who were then living at Bromham, near Devizes.

My mother and aunt (two of the most admirable
of women) loved each other most tenderly and
truly.

In the following winter, 1826-7, in what month I
do not remember—occurred the first separation that
had taken place in our family since the death of my
little sister. My youngest brother, Hanmer, just
thirteen years of age, was taken to Portsmouth to
enter as a midshipman on board the Asia, under the
command of Sir Edward Codrington. It must have
cost my mother a pang, loving her children as she

did, to send one of them, so young, into an adven- 1826
turous and hazardous profession.

Yet perhaps it was less difficult to her, than it
might have been to some mothers ; at least she
must have been somewhat prepared for it, for she
had always cherished such a feeling of intense love
and admiration (I might almost say adoration) for
the two heroic professions, the military and naval ;
she had been so much used in talking with her children, to expatiate on the glories of those professions,
and to hold up distinguished soldiers and sailors as
models for imitation, that she must certainly have
been prepared for the practical result of her lessons.

Moreover Hanmer was going to sea under favourable auspices, as Henry Codrington, an intimate
friend, was a lieutenant on board the same ship, and
there seemed to be no threatening danger, for no one
dreamed of war as likely to be near. My reading
during these years, 1826 and 1827, was by no means
confined to the subjects which I studied under Mr.
Matthews, nor yet to my dearly loved botany. I
have mentioned before my early love for Walter
Scott's poetry ; a love which proved very constant,
so that even at the present day I take a degree of
pleasure in those poems which is (I think) shared
by few of my contemporaries. It was not till the
year 1826, that I made acquaintance with Byron's
poetry :—the " Bride of Abydos," the " Giaour," the
" Corsair," " Lara," were first read aloud to us (my
brother and myself) by our father, and afterwards
I read and re-read them with ever increasing delight.

1826 " Childe Harold " I did not learn to appreciate till a little later.

Thomas Moore about the same time became another of my chief poetical favourites. I delighted in " Lalla Rookh." Campbell also I read with pleasure. Of course I read Scott's novels as fast as they came out, and enjoyed them heartily.

Another novelist—or romance writer—whom we all read with delight in those days was Fenimore Cooper, the American. I hardly know whether his naval stories or those of which Red Indians were the heroes, were the most admired : my mother was enthusiastic about both and quite carried me along with her.

It is very many years—certainly above forty— since I have looked into any of these works of Cooper ; and I feel considerable doubt how far I should relish them now. Walter Scott I read over and over again with fresh pleasure. But though I own that my taste in fiction in those days inclined decidedly to the romantic, yet I do remember that I read Miss Austen with great pleasure, though perhaps hardly with so much enjoyment as I have experienced since.

At the same time we had plenty of graver reading. Both Edward and I, in the years of which I speak, read fairly through Gibbon's " Decline and Fall." There was a great deficiency in our supply of historical knowledge ; to instruct us in the earlier history of Rome, to bring us down to the time at which Gibbon begins, we had nothing but Hooke and

Goldsmith; the one very dull, the other excessively 1827
superficial. For Grecian history we read Mitford—
very clever, but utterly one-sided and untrustworthy.

A more recent work, and a very important and
valuable one, which I read, by my father's desire,
during this time of quiet and of intellectual activity,
was Hallam's "Middle Ages." It had not long been
published, and it had made a great impression on
my father, who read many portions of it aloud to my
mother and me. He had been especially struck by
some magnificent passages relating to Italy.

I do not know how long the state of my mother's
health had excited the anxiety of my father before
I was aware of his uneasiness, I well remember
that it was in the summer of 1827—I think in June
—that he first told me of it, or at least of the effect
it produced on his plans. We were on our way
from Barton to Southwold, whither my mother had
been advised to go for a time, for the benefit of the
sea air; and while the horses baited at Botesdale,
my father and I took a walk in the lanes of the
neighbourhood. It was then that he first told me
that the state of my mother's health had for some
time made him very uneasy, that she seemed to
grow worse instead of better; and that he had at
last made up his mind to go abroad with her and
us three boys before the winter of that year.

It was, I think, in the same walk that my father
first (as well as I remember) began to talk to me
about his political views and objects. He explained
his political creed, which was that of a wise, liberal-

1827 minded, and reasonable Whig, distinct alike from
Toryism and Democracy. But he went on to say
that, even if my mother's health had not compelled
him to go abroad, he had no serious thoughts of
entering Parliament, or political life at that time;
for there was nothing that he perceived in the
political state of the time, calculated to excite any
warm feeling of party spirit, or to call for any
patriotic zeal. He added, however, that this state
of tranquillity was not likely to be durable, and that
it might not be long before he and other men who
took a warm interest in the welfare of the country,
might feel themselves called upon to proclaim
openly their views on public affairs.

In fact that summer was not over before the
unexpected death of Canning put an end to the
delusive calm. I hardly remember to have seen my
father more shocked and agitated by any political
event than he was by the death of George Canning.
He, of course, put no trust in the ephemeral Ministry of Lord Goderich, and assuming (as he naturally
might), that the government of Wellington, and
Peel, would be founded on principles of political
and religious intolerance, he looked forward to hard
and difficult struggles between the contending principles.

When we set out from Barton for our foreign tour
(from which my dear mother was never to return), I
was of the age of eighteen years and about eight
months.

I must say of my youthful days, as Gibbon says—

"I seldom mounted a horse, I never handled a gun." 1827

We (that is to say my father and mother, my brothers, Edward and Henry, and myself, with I do not exactly remember how many servants) landed at Calais on Saturday, October 13th, and spent the night there. This was, for my brothers and myself, our first introduction—I might almost say to a new world—so new and so strange did almost everything appear which met our eyes and our ears, on first landing on the other side of the Channel. Even at the present day, when so many years of continual peaceful communication between the two countries have done much to smooth down the differences, and when, above all, the sameness of railway travelling has done so much in the same way, even now one cannot cross over to Calais or Boulogne without being struck with many differences between the ways of the French and our ways. In 1827 the contrast was much more striking. Most especially we were impressed by it, when, on the following morning, we saw the *turn out* for the commencement of our journey by post to Paris. The rough, strong, but unkempt-looking horses, the harness of long ropes, leaving what seemed an unreasonable distance between the horses and the carriage; above all, the huge jack-boots and the huge whips of the postillions; all these peculiarities, so strongly marked out from what we had ever seen in England, filled us with wonder and amusement.

We were still more amused when in proceeding along the road, we heard the noise (I might almost

1827 say the tune) which the postillions made by the crack of their great whips whenever we approached a post house or rattled through the streets of a village or town.

By reason both of my mother's weak health and of the late season, and consequently short daylight, we travelled very slowly; we left Calais on October 14th, slept that night at Boulogne; the 15th at Abbeville; the 16th at Amiens; the 17th at Chantilly; and reached Paris on the 18th.

My father paid much attention to the architecture of the several churches we passed, and especially to the truly magnificent cathedral of Amiens. My studies had not yet been directed to those subjects. I kept my eyes open, of course, to everything I could meet with in the way of my favourite science; but at that season of the year there were very few Botanical curiosities to be seen. The first plant I gathered on foreign ground was the " Sea Buckthorn," *Hippophœ rhamnoides*, which grew in plenty on the sandhills of the shore near Calais; this, indeed, is a native English plant, but very local, and I had never before seen it except in gardens.

The next novelty I met with was the Field Eryngo, *Eryngium campestre*, growing near Boulogne; this is a very common plant throughout France and all the South of Europe, but very rare in England, indeed a rather doubtful native. The only other foreign plant which I observed before our arrival at Paris was the wild Marigold, *Calendula arvensis*.

We remained at Paris from the 18th to the 27th

of October, and employed our time pretty diligently 1827
in sight-seeing, for the French capital was as new
to my father as to his sons. He was very much
interested and delighted by the numerous and
striking monuments of taste, magnificence and
utility, with which Napoleon the First had enriched
his capital; and in his journal my father expatiated
at length on these works :—the Bridges, the Quays
the Bourse, the wide streets, the immense magazines
for grain and for wine, the Abattoirs, the Reservoir
for supplying the city with water, and various other
great works for ornament and use. In such matters
Paris in those days appeared to have greatly the
advantage over London. We were much interested
by the Cemetery of Père la Chaise, which contained
the monuments of so many famous men. In my
journal, after mentioning the monuments of Massena
and Davoust, I particularly noticed the absence of
any record of Ney, his burial place being marked
only by a simple enclosure of palissades, without
any inscription. Our *lacquais de place*, however, was
eager to supply this deficiency of information, both
as to Ney and his fellow victim, Labedoyère.

I remember that at the time at which I speak,
and for some years before and after, I was disposed
to be a much more zealous admirer of Napoleon
than I have ever been since.

His astonishing career was then comparatively
recent ; the overpowering impressions it had left on
men's minds, causing some almost to view him as a
demi-god, and others to denounce him almost as a

D

1827 demon, were still fresh and vivid. I had latterly been a good deal under the influence of my Uncle William, who was one of the absolute worshippers of Napoleon ; and though my father was much more sane and temperate he could not fail to be greatly interested in the achievements of the great Corsican, and gratified by beholding the recent traces of his power and genius. I had read and heard enough of the recent history of Europe to understand why my father admired Napoleon and, as was natural in a lad of eighteen, I rushed headlong from admiration into enthusiasm. I may add that what, in this visit to Paris, we saw or heard of the doings of the restored Bourbons, was in no way calculated to correct a tendency to Bonapartism.

In this stay at Paris, we repeatedly visited the Jardins des Plantes, and had much pleasure in seeing it ; though the acquaintance I made with it this time was very slight compared with that of later years. What interested us most at this time was the collection of living animals, which, widely as it has since been surpassed by our Zoological Gardens, was yet far superior to any menagerie which I had before seen. It was a pleasant novelty to see the animals so (comparatively) well lodged as they were here. The great attraction was the Giraffe—the first, I believe, which had been seen alive in Europe since the sixteenth century, and which, naturally, had excited a vast deal of curiosity and interest in Paris.

I find by my journal that I was not much struck

by the collection of living plants in the greenhouses 1827
of the Jardins des Plantes; that I thought the
collection of minerals inferior to that in the British
Museum; but I greatly admired the zoological department, especially the stuffed quadrupeds. We
left Paris on the 27th October, and crossed the river
Var (at that time the frontier between France and
Italy), on the third of December. This was a
tedious and fatiguing journey, especially for my dear
mother, who was in bad health, and ill-qualified to
bear the trials of such rough travelling. We travelled
post, and our carriages were heavy. The roads for
the most part very bad, the posting very slow, and
the season being winter, and the days consequently
short, so also were our days' journeys, and we were
forced to spend much time at inns, which were
generally very comfortless.

For a long way from Paris towards Autun, I was
very much struck with the want of beauty in the
scenery, the scarcity of trees, and the monotony of
the vine-covered hills.

At Autun, the bold and steep range of hills behind
the town afforded us some amusement; and I met
with a few interesting plants, such as *Genista pilosa*,
and the spotted-leaved variety of *Hieracium murorum*.
It was near Tournus, between Châlons and Mâcon,
that we first came into pretty country, and it went on
improving as we went southward, till in the valley of
the Rhone below Lyons we found a rapid succession
of beautiful and interesting country.

From the windows of our inn at Mâcon, I had my

1827 first view of the Alps: with a telescope we could distinctly make out Mont Blanc and some other peaks. The first view of the Alps may be considered as an event in one's life—especially in that of one brought up in a flat country—as the first view of the sea is to the inhabitant of an inland district. In the next day's journey, however, from Lyons to Vienne (November 9th) I was much struck by the beauty of the scenery about Vienne, and almost as much by the ancient Roman temple which we saw in that town. This was the first ancient Roman Temple which I remember to have seen, and it interested me very much

Proceeding southward, by Valence, Montelimart and Pont St. Esprit to Nismes, we were in an almost continual state of delight with the beauty of the scenery, which was unlike anything I had seen before. I have since made that journey (from Lyons to Avignon or Nismes) three or four times, sometimes from North to South, sometimes in the opposite direction ; and every time, in spite of the terrible *Vent de Bise,* I have heartily admired the Rhone and its valley. It was about Montelimart, or at least between Valence and Avignon, that I first became acquainted with the real native vegetation of South Europe—the Mediterranean Flora. While the horses were changed at Derbieres (between Valence and Montelimart) my father and I ascended a very steep and remarkably arid hill behind the town, and I was immediately struck with the peculiarity of its vegetation ; it was clothed, rather thinly, with Box,

Juniper, stunted Evergreen Oaks, and a greyish 1827
thorny shrub allied to Cytisus, amidst which grew
abundance of Lavender, *Teucrium Polium*, *Plantago
Cynops*, and several other plants which were new to
me. There was scarcely any grass of such kinds as
I had been used to, but I saw there for the first time
the *Cynodon Dactylon* and *Andropogon Ischæmum*,
two of the grasses characteristic of warm countries.

Before reaching Avignon we made a deviation
from the direct road in order to see, first the famous
Pont de Gard, and then, the antiquities of Nismes.
I hardly need say, that we found them well worthy
of a greater effort and a greater sacrifice of time.
We were all delighted with the beauty and grandeur
of the Roman aqueduct itself, and almost more still
with the remarkable seenery amidst which it is
situated. Here too, my botanical zeal was highly
excited. I never previously saw in any one place
such a multitude of curious and interesting plants as
among the thickets and warm sunny rocks (screened
from the violence of the Mistral) down by the river,
just below the " Pont du Gard." The vegetation
was in fact that which is especially characteristic of
the *Mediterranean* region : here I first saw, growing
wild, the *Similax aspera*, *Asparagus acutifolius*, *Jasmi-
num fruticans*, the *Paliurus*, and the *Adiatum*. I
saw likewise at the same place, that very curious
insect, the *Praying Mantis*. We remained at Nismes
the whole of the 14th and part of the 15th, finding
plenty to occupy our attention and excite our interest.
We were delighted especially with the amphitheatre

1827 (the first which I and my brothers had ever seen)
and with the beautiful temple popularly called the
Maison carrée.

My father found much also to admire in the
modern city of Nismes, and in the improvements
which had been executed, or were going on within
or around it.

It was while at Nismes that we first heard of the
battle of Navarino.   This naturally caused my
father and mother much anxiety, for they knew that
their son Hanmer, being on board the British
Admiral's ship, must have been in the thick of the
fight, and exposed to much danger.   It was not till
arriving at Marseilles on the 17th, that they received letters with the information that poor Hanmer
had lost his arm in the battle, while behaving with
so much courage as to earn high commendation from
his superior officers.   This painful news made some
change in our movements ; my father was induced to
halt for three days at Marseilles in expectation of the
Malta packet, which was expected to bring more
news of poor Hanmer ;—but it did not arrive.   Our
next move was to Aix (the old capital of Provence)
on our way to Nice, but at Aix my father fell dangerously ill, whether in consequence of mental
agitation or of the dreadful smells from the harbour
of Marseilles, and we were obliged to remain stationary (fortunately in good quarters) until the first of
December, when my father was sufficiently recovered
to continue his journey.   On the 2nd we reached
Frejus (the ancient Forum Julii), the birth-place of

Agricola, and celebrated in modern times as the 1827
place where Napoleon landed on his return from
Egypt, and embarked when banished to Elba. On
the 3rd we went over the Lestrelles to Cannes, a most
beautiful country; a little beyond Cannes the road
proceeds along the margin of a pretty bay where
Napoleon landed on his return from Elba, and in
which was lying the frigate La Syréne, on board of
which Admiral de Rigny was at the battle of Navarino.
From thence we went on to Antibes, the last post
in France, crossed the river Var, which separates the
French dominions from those of the King of Sardinia,
and reached Nice late at night. We remained here
till the 9th and were very much pleased with the
town and the scenery.

(I continue to omit a great part of this journal
which gives a description of the scenery, natural
history, and art of the places they visited.—F. J. B.)

I never saw elsewhere so great a number of priests
—one meets them at every turn, and their black
dresses and queer hats are very ungraceful. The
people of the lower class here are very swarthy in
complexion, "the colour of mahogany", but often
handsome, and the dresses of the women are pretty
and picturesque, especially the nets of various
colours they often wear on their heads, though the
most usual coiffure is a coloured handkerchief tied
over the head and under the chin.

On the 10th we left Nice on our way to Genoa, not
by post as we had hitherto travelled, but with
vetturini, who supplied us with two horses to the

1827 open and four to the closed carriage, and rode themselves. We rested for some time to bait the horses
at Mentone, an inconsiderable place, but with a
pretty good inn ; then proceeded to Ventimiglia, our
destination for the night. The next morning we
set off early, baited at San Remo, which, when
Genoa was a republic, was one of the most considerable places in its territory ; and spent the night at
Oneglia. On the 13th we passed Loano, remarkable for a battle fought in the mountains near it in
1795, in which the Austrians commanded by Argen-
teau and Wailis were defeated by the French under
Massena and Schérer.

We arrived at Genoa on the 13th of December,
and remained there four months. While there Sir
Edward Codrington sent Hanmer to stay with us to
recover from the amputation of his arm.

1828

LETTERS.

Villa Majon, Genoa.
March 31st, 1828.

DEAR AUNT EMILY*

How can I possibly make an adequate apology for my long silence? I have not written to you since we were at Lyons and that, I am afraid, is more than four months ago.

I cannot find any excuse to offer except my dilatory habits, and therefore I believe the best thing I can do is to acknowledge my fault, and throw myself on the mercy of the court. Having done this, the next thing is to attempt to make reparation as far as I can, by giving you some little information (such as it is in my power to communicate) relative to the botany of this region. As for sending you a list of all the plants which I have observed since we entered Italy, it is out of the question, for such a list would occupy as much space as a moderate-sized letter. The environs of Genoa are extremely hilly, presenting in fact scarcely any plain ground but a succession of hills, variously grouped and increasing in height from the seashore up to the main ridge of the Appennines. The less considerable of these are

* His cousin, Miss Napier, afterwards his father's second wife,

1828 covered with olive-trees, as indeed are the lower
parts of all, the rocky soil being formed into a succession of terraces, which are richly cultivated, so
that corn appears under the olive-trees, and vines
frequently along the margins of the terraces.   Higher
up, where cultivation ceases, the steep rocky sides
of the mountains are rich in a variety of beautiful
wild shrubs, most of which are favourite ornaments
of our English gardens :—

Arbutus, Myrtle, Erica arborea, Cistus salvifolius,
Alaternus, and Ilex, are very abundant, and in some
sheltered nooks and hollows attain a considerable
size ; the Mastic (Pistacia Lantiscus) is less common.
The Erica arborea is now in full bloom, and has a
very beautiful appearance, being from two to four feet
high, very much branched, somewhat pyramidal in
its general form, and covered with a profusion of
delicate white flowers.   Of trees there is certainly a
considerable scarcity ; chestnut is almost the only
sort that grows on the mountains, and of this there
is no great quantity, except in some particular places.
With respect to herbaceous plants, I am under some
difficulty, for it would be impossible to give you even
a brief account of the species which are new to me,
and a mere list of names would probably be neither
amusing nor instructive.   However, I cannot omit
the beautiful tribe of Orchideous plants, in which
this neighbourhood, even so early as the present
month, appears to be peculiarly rich.   I have already
found three species of Serapias, two of Orchis, and
two of Ophrys, all growing on the open steep sunny

sides of the mountains, among short grass. Of the 1828 two species of Ophrys, one is O. aranifera, the Spider Ophrys, which I suppose you know by description, as it is a British plant, though a rare one; here it is very common, and the earliest of the tribe, flowering before the end of January—the other is O. arachnites, almost intermediate between the Spider and the Bee Orchis, having pretty nearly the lip of the former, though more variegated, with pink petals like the latter. One of the Orchises is our common English O. Morio,—the other appears to me to be a cream coloured variety of O. mascula, though some authors have considered it a distinct species. But the three species of Serapias are the most worthy of notice, being all beautiful, all remarkable in form and colour, and all strangers to Britain,—I cannot resist giving you a brief sketch of them. The first and handsomest of the three, Serapias cordigera, has large flowers, five or six in a spike, with the petals united very firmly into a kind of hood of a purplish colour, and the lip of the nectary divided into three lobes, of which the two side ones are erect and rounded, while the central is heart-shaped, pointed, pendulous, and very hairy in the upper part; colour varying from a deep brownish red to a purplish rust-colour. This species is abundant in several places near Genoa, and is one of the most singular and handsome plants of its tribe that I have seen. The second, S. Lingua, flowers later, is a more slender plant, with very much smaller flowers, though similar in form; the lip is deep red, almost crimson,

1828 but quite white towards the base, and is scarcely at
all hairy. The third, Serapias triloba, is less remarkable in appearance, the lip being in three blunt,
equal spreading lobes; quite smooth, and of a pale
rosy purple; this however is much the most rare of
the three, in fact I have only found a single specimen of it; while the other two seem to be not uncommon here. These three are altogether singularly interesting and attractive plants, but unfortunately, like most of the Orchis tribe, lose their
beauty entirely in a dried state, so that except as
memoranda it is scarcely worth while to preserve
specimens. I have therefore carefully made coloured
drawings, which I hope some time or other to have
an opportunity of shewing to you.—I have moreover
found here some plants which are natives of Britain,
though of very rare occurrence there,—such as
Pulmonaria officinalis, Symphytum tuberosum,
Hyacinthus racemosus, Crocus vernus, and Helle-
borus viridus. It is remarkable that some plants
which in England are natives of low meadows, grow
here at a considerable elevation on the dry rocky
parched sides of the mountains, this is the case in
particular with Orchis Morio, Crocus vernus, and
the common cowslip.

Among the most common wild flowers here are
Anemone hortensis, Anemone Hepatica, and Globu-
laria vulgaris,—this last a very pretty little thing with
delicate blue flowers. Under walls and in narrow
stony lanes, grows a small species of Arum (A. Aris-
æma), the spatha of which is curiously marked with

regular longitudinal stripes of brown or dark green. 1828
There are very many other plants which I would
describe if I had room, and if I thought that it
could be amusing to you.

My father and I are just at present alone, the rest
of the party being all gone on a trip to Spezia,
aboard the Gulnare, and having now been absent
five days, although the wind has been very fair. I
fancy the extreme fineness of the weather has
tempted them to prolong their cruise, and perhaps
Colonel Light, having once got his vessel out of the
harbour of Genoa, is inclined to give her and his
men some exercise, before he returns. Hanmer left
us at the same time, on board the frigate Rattle-
snake, to return to Malta, after having been with us
nearly three months,—a time very pleasant to us
and I believe to him also.

He seems to have thriven upon his Navarino
adventure, being in excellent health, and far merrier
and more constantly in good spirits than he was
before he left England; and I never saw any one
more attached to his profession, or more proud of it.
It does put me out of all patience to see the way in
which some of the English newspapers speak of the
battle of Navarino, and their utterly unfounded
attacks upon Sir Edward Codrington,—while at the
same time I observe that even the most violent
Tories expressed themselves with great caution
upon the subject in Parliament. To leave this
digression;—I will fairly own that Genoa at first
disappointed me much, but after the first week or

1828 two I began to like it, and it has continued improving and gaining upon my affections more and more every week, insomuch that I shall be really sorry to leave it, which we must do in about a fortnight, to make a tour in Tuscany, though I am in hopes that we may return hither for the summer.   Our stay has been rendered peculiarly pleasant by Hanmer having been with us so long,   The place itself also has many attractions, and indeed I have not seen any book of travels that does it justice.   The palaces are on the greatest scale of magnificence,—the collections of pictures numerous and good,—the climate delightful,—the society agreeable,—and the general view of the city and bay is of a degree of beauty which pleases one more and more every time one sees it.   I hope that your love of botany is not abated, and that you continue to study with success the plants of Dawlish—although I have not heard a word, good or bad, on the subject since I left England.   How do the mosses go on ?   Pray re-member me to Sarah and Cissy.

Yours very sincerely,

C. J. F. Bunbury.

MEMOIRS.

On the 19th of April Sir Henry and Lady Bun- 1828
bury and their three eldest sons travelled by La
Spezia, Carrara, Lucca, Pisa, and Florence, to
Sienna ; where they spent ten days with Captain
and Mrs. Henry Napier very agreeably. From
thence they visited Florence, Bologna, Parma,
Modena, and Milan, and returned by the Lake of
Como to Genoa, where they arrived about the
middle of July.

Charles Bunbury, who was just nineteen, wrote a
journal of his tour, which I have in my possession.
In this he gives a full account of the various works
of art (pictures, sculpture and architecture), also of
the museums of Natural History. He gives a very
full account of the botany of the country they pass
through, also of its geology, and speaks with
much rapture of the beauty of the scenery. His
father had taken a large house in Genoa, Villa
Mari, finely situated on the hill-side above the city ;
but his mother's health had become worse, and it
was thought advisable to remove her for a time to
Nervi, a small town on the coast, a few miles east of
Genoa. Here she died on the 15th of September
very suddenly. There had been so little apprehen-
sion of her immediate danger, that her husband, to

1828 whom she was dearer than anything else on earth, was absent on a short excursion he was making with his son Edward through the Alps near Turin, so that at the time of her death she was alone with her sons Charles and Henry. I will quote Charles' own words of his mother which he wrote in his memoir of his father.

" I might be tempted to dwell for some time on my mother's character, but I feel that it is utterly out of my power to do justice to her by anything I can say. It is indeed impossible for me to imagine a better mother. Hasty observers might have thought her too fond, too indulgent ; but, with all this indulgence, her watch fulness was unceasing, and no symptom of selfishness, or any other moral evil, could escape her notice and reproof. She delighted in teaching her children, and taught them so pleasantly, that the memory of those early instructions is still treasured among their dearest remembrances. As I write, the image of her dressing room, in which we used to sit and draw and read with her, rises up in my mind. It was not only, or mainly, by regular lessons that she taught us, but by watching, encouraging, and guiding our intellectual tendencies, and fostering every indication of taste or talent. What she was as a wife, will perhaps be sufficiently shown by a passage in one of my father's letters to her brother, Henry Stephen Fox, soon after her death.

He writes thus,—" My existence has been so completely identified with my poor Louisa, that I cannot clear my head from the strange confused impressions,

which make me feel it almost impossible that I 1828
should be alive and she should be no more. I believe
there never were two persons who had come to rely
for there comfort and tranquillity of mind upon their
mutual affection and confidence more than your
sister and myself."

Charles Bunbury goes on to say, "she had a well
cultivated mind, and delighted in poetry. In literature as in ordinary life, her taste was much guided by
her moral feelings, and her feelings were always
pure, generous, and noble. No one had a more
generous indignation against everything mean, sordid
or cruel; no one had a more tender charity and
compassion for the weak, the erring, and the unfortunate. Her piety was deep, earnest, and unwaver-
ing, but it was eminently cheerful; her religion was
as free from gloom and austerity as from intolerance.
In the month of October my father made a short
tour in the Tyrol, taking with him his three elder
sons, hoping that the fine scenery and the variety of
new objects might benefit his health, both of body
and mind."

Of this tour Charles Bunbury wrote a very interesting journal, with a detailed account of its scenery,
geology and botany of what he saw. He also alludes
much in his journal to Napoleon's Campaigns, as
they went over much of the same ground.

On their return to Genoa they found Henry
Stephen Fox, his mother's brother, whom they had
long expected, arrived there, and established in the
Villa Mari. He was at that time Secretary of Lega-

1829 tion at Naples. His visit had been promised and
expected for months before, but although he had
been informed of the illness of his sister, whom he
had not seen for many years, he was, perhaps, not
aware of her danger, and had thus, with his usual
habit of procrastination, delayed his coming until it
was too late. He was a man whose talents promised
him a brilliant career; but neutralized by a want
of concentration, and of firmness of purpose,
they only served to render him a delightful
companion.

He remained at Villa Mari with his brother-in-law
and nephews, until the end of the winter, when he
set out before them for Rome. Soon after Sir Henry
broke up his establishment at Genoa, and he and
his sons travelled to Florence, and thence to Rome
by the beautiful and interesting route of Arezzo,
Cortona, Perugia, Spoleto, Terni, and Narni. They
arrived at Rome about the end of February, and remained there about two months. Charles Bunbury
describes in his Journal the antiquities, the works
of art, and the beautiful scenery of that most
interesting city. Leaving Rome in the last days of
April they crossed the Apennines from Foligno to
Macerata and Ancona, and proceeded by the line of
the old Flaminian and Æmilian roads, by Pesaro and
Rimini to Bologna, making, however, a deviation
from the direct route to visit that most curious and
interesting old city, Ravenna. Thence they went
by Padua to Venice, and after a stay of ten delightful
days in that city, visited the wonderfully fine moun-

tain scenery of the Alps, to the north of it. They 1830
returned to Genoa on the first of June.

In the month of September Sir Henry returned to
England with his two eldest sons, Henry Bunbury
having been left at Zurich to study French and
German, and military drawing.

In October, 1829, Charles and Edward Bunbury
entered Trinity College, Cambridge.

LETTERS.

Trinity College,
Tuesday, March 16th, 1830.

MY DEAR FATHER,

I suppose you are waiting to hear from us before
you write, and perhaps you will call me (as you
once did before) an abominable bumble-bee, for not
having written sooner, but I have put it off from day
to day in the hope of having something interesting
to tell you. I went to the Cambridge ball last
Wednesday, and liked it very well. The number of
Cantabs at the ball was not so great as I expected;
there were some of our best Trinity men, Milnes,*
Arundel Fitzroy,—and some on the other hand who
were most curious specimens of Cambridge men. I
stayed till the end of the ball, and did not get back

* Afterwards Lord Houghton.

1830 to my lodgings before three o'clock, which is a thing
which has never happened to me before at Cambridge.

I have retrieved my *Union* reputation by a speech
last Tuesday for which I have been very much com-
plimented ; it was certainly much better than any I
had made before, because I had fully arranged my
ideas and knew what I wanted to say, and then the
words came of themselves.   We had a debate yesterday on the Union with Ireland, in which Edward made
his first speech, and succeeded extremely well—a
hundred times better than I did in my first attempt.
He and one other man spoke in favour of the Union
—five against it ; yet when we came to a division, the
majority against it was only eight.   Among the
speakers on that side was Mr. Matthew's friend
Praed,* who was a distinguished member of the
Union, while he was an Undergraduate—he made the
best speech I heard that evening.   What a confusion
I have made between the two unions !   I hope you
will be able to find out my meaning.

Last Sunday we breakfasted with O'Brien,† who is
a very agreeable clever fellow, and after breakfast a
party of us (consisting of O'Brien, Milnes, Grey,
Cavendish. Duncan, Fitzroy, Edward and I) set out
for a rambling walk across the country, over hedges
and ditches, the active ones of the party leaping
them, and the rest scrambling over as they could.

---

* Winthorp Praed.

† A very intimate college friend of Charles Bunbury's.

Afterwards we went to St. Mary's to hear the ser- 1830
mon, and then paraded about the college walks until
dinner time—and a tea-party at Fitzroy's rooms
finished the day. You cannot think how much more
I have enjoyed this last month than any previous
part of my "academical career," and I daresay the
summer term will be very pleasant, but I shall then
have to make a new arrangement of my time, as
there will not be the long evenings which have
hitherto afforded me so much reading time. With
Mr. Coddington we have gone through Trigonometry,
and entered upon one of the second year's subjects,
the Theory of Equations, which as far as we have
gone yet, is not very difficult.

They have chosen a new Vice-master, who rejoices
in the name of John Brown; the only remarkable
thing I have yet heard of him, is that it is very
likely he may change the dinner hour of the College.
Milnes is going after all to take his degree quietly
this term, instead of publishing a pamphlet, but he
talks of coming up again next term to make a fare-
well address to the Union. He will really be a loss
here, in spite of his conceit and his oddity I like
him, he is so thoroughly good natured, and so amusing. I hear he is going to Mr. Wyvill's in the
vacation; I should like to know what he and Mr.
Matthews* will think of each other, for I cannot
imagine two characters more totally different.

I wrote to my aunt on Saturday. I hope I shall

---

* Their former tutor, who went to Mr. Wyvill's.

1830 soon get a letter from you, and hear that you are quite well, and Mrs. Gwyn the same. We have not yet been much plagued by the east winds; I do not know how they may have been in London.

Ever your affectionate son,

C. J. F. BUNBURY.

———

From Sir Henry Bunbury to his sons Charles and Edward.

June 3rd, 1830.

MY DEAR BOYS,

The first and fondest wish of my heart is to live in perfect confidence with you: my greatest blessing is your affection, it is painful for me to withhold from your knowledge anything which concerns me materially; and still more so if it be a matter in which your own comfort or interests may be involved. I have therefore made up my mind to confide to you a matter which must still be considered as a secret, and which indeed I am hardly at liberty to mention on account of other persons ; but I have that confidence in your discretion, as well as in your love for me, that I will not conceal from you the circumstances in which I stand, or the event to which I look forward.

It is probable that before the next winter I shall be married. I know that the idea of a stepmother is associated in the minds of young persons with 1830 painful feelings and apprehensions, and I should hate myself if I had not considered the subject, long and anxiously, as it regarded your comfort and advantage. I have pondered it over and over again, and I can say conscientiously that I have weighed it more carefully with regard to you than to myself. But you, at your time of life, can hardly form an adequate idea of the sad change in my condition which the loss of your dear, dear mother has occasioned. You cannot know what it is, after twenty-two years of love and companionship, and entire and mutual confidence, to find myself, as the autumn of life draws on, a solitary being in my once cheerful and happy home. I can never love any other woman as I have loved your dear mother. The memory of our early days, the strong bonds woven by our common interests in our children, and the long habits of mutual affection and respect, and intimate confidence, united us by ties which can never be formed again. What I feel the gnawing want of, is a tender friend, a constant companion, a being to whose affection and good sense I may confide my cares, and to whom I may look for counsel and assistance. But besides this I want, for your sakes, a cheerful *home*, to which you may always turn for pleasant society, rational amusement and quiet comfort. While I remain solitary and uneasy, and moody, your home must be sadly destitute of attraction. *I am thoroughly satisfied* that the restoration of cheerfulness and society to this house, is

1830 one of the best services that I can render to you and your brothers. I could say much more to strengthen the views I have confided to you, both as the subject regards your comfort, and my own; but you will be impatient to learn who the person is, who will probably share your father's cares and pleasures through the rest of his life. It is one who loves you both dearly, and who is fully capable of estimating the qualities of your minds and hearts. It is Emily Napier. I have been intimately acquainted with the strength of her understanding and the nobleness of her heart for twenty years. I anticipate your viewing this prospect before us with pleasure; and I am sure you will feel convinced that I have not swerved for one moment from the main object of my life, the promoting of your happiness. God Almighty bless you.

I am, your ever affectionate,

Father.

In the summer of 1830 Sir Henry Bunbury was elected member for the county of Suffolk, and soon after the Suffolk election he set out for Pau in the south of France, where he was married on September 22nd to Miss Emily Louisa Augusta Napier, the only surviving daughter of Colonel the Honour-

able George Napier and Lady Sarah Napier. She 1830
was first cousin once removed, to Charles Bunbury's
mother; the one being grand-daughter of Lady
Holland, the eldest daughter of the second Duke of
Richmond ; the other, daughter of Lady Sarah, who
was the youngest daughter of the same Duke. Miss
Napier at the time of her marriage was residing with
her brother George at Pau. It was a marriage pro-
ductive of much happiness, both to Sir Henry and
his sons. His two nieces Sarah and Cecilia Napier
were dear to these sons as sisters ; the youngest of
these sisters became in after years the wife of Charles
Bunbury's brother Henry.

In the autumn of this year, Charles Bunbury set
off for a tour in Scotland with his brother Edward.

Edinburgh.
Monday, August 30th, 1830.

My Dear Father,

We arrived here safe and sound on Friday afternoon, having come in twenty-two hours by the mail
from York, where we spent one night, and the
greater part of a day. We found nothing to see
at York except the cathedral, which is certainly one
of the grandest and most beautiful buildings I ever
saw ; the interior does not appear to advantage at
present, as the repairs of the part that was burnt
are going on, and the choir is full of scaffolding and

1830 rubbish; but in spite of these disadvantages the view up the nave is uncommonly beautiful. There is much fine old painted glass, and this circumstance adds very much to the fine effect of the interior. The two great towers of the west front struck me as singularly beautiful, having great lightness in their general effect, together with extreme richness in the details, and the large window in the same front has a beautiful leafy tracery. Being impatient to get on to Edinburgh, we resolved not to stop at Durham, and as the coach unluckily passed through it in the night, we saw nothing of it. It was raining hard when we reached Edinburgh. The next day we called on Mr. Hasted, who assisted us in arranging a plan for our highland tour, and went with us to the college museum, where we spent some hours very pleasantly. There is a superb collection of birds (said to be one of the finest in Europe), some good minerals, and a great number of quadrupeds, very well stuffed—among them a walrus, which is an animal I never saw before. Yesterday was a tolerable day, though showery, and this morning has been very fine, consequently we have seen a good deal of the town, and have already been up to the top of Arthur's Seat. I certainly never saw a more picturesque town than Edinburgh—it has far surpassed my expectations, and Edward is quite in raptures with it. In many points it reminds us of Genoa, particularly in the steep streets, the excessively high old houses (some of them of nine or ten stories), and the bridges connecting different streets, from which one looks down upon the roofs of the 1830 houses in the old town. Indeed it has altogether the appearance of a foreign city, and the picturesqueness and singularity of its appearance are delightful after the dull prosaic towns that one is used to in England. The castle, standing on the brow of a bold precipitous hill overlooking the town, has a most imposing and majestic appearance, and the hill called Arthur's Seat, which is to the east of the town, though only 820 feet high, is so steep and rocky that it has completely a mountain character, and yesterday morning when we saw it through the haze, it was difficult not to imagine it a high mountain.

Salisbury Crags, which form part of the same group of hills with Arthur's Seat, are very picturesque in themselves, and the view of Edinburgh from a path along the face of them is most beautiful. In one place, near the foot of Arthur's Seat, there is some columnar basalt, but not remarkably distinct. We have sketched out a very nice plan for our tour, by which, in the space of a fortnight, we hope to see Blair Athol, Loch Tay, Loch Katrine, and Loch Lomond. Weather allowing, we shall leave this place on Wednesday.

Pray give my love to our friends at Pau, and believe me ever your very affectionate son,

C. J. F. BUNBURY.

1831   Charles Bunbury while at Cambridge did not
neglect his favourite study of natural history ; he
wrote in " Lindley's Introduction to Botany," published 1830, as follows :—

" I first read this book when I was a student at
Cambridge in 1831, the same year that I read
Lyell's Principles of Geology (the first volume at
least), and the two books made as deep an impression
on me as any books I ever read, and gave me entirely new views and ideas of the whole of their
respective sciences.   Till then I was almost entirely
a Linnean botanist.   C. B."

Charles Lyell's book was also published in 1830.

F. J. Bunbury.

Trinity College.
Sunday, April 17th, 1831.

My Dear Father,

If I do not grow conceited, I am sure it will not be
the fault of the Dons, for they have been loading me
with compliments ever since I came up.   But
seriously, it is a great satisfaction to find that what
we did in the examination was so much approved of,
and that we are considered likely to take very good
degrees, in classics as well as mathematics, if we
persevere.   Mr. Peacock and Mr. Evans have both

strongly recommended me to read for the Classical 1831
Tripos, which I had before had no thoughts of
doing ; but with such encouragement I shall certainly
set to work in earnest, though rather late, as I have
now less than two years before me. I am in hopes,
from what I have heard concerning this examination,
that Edward has shown himself not inferior in
classics to the best of his contemporaries.

Mr. Peacock gave much praise to the style of my
translation, and said that was a point in which those
who had been brought up at home were particularly
apt to fail ; he told me also that he did not remember
another instance of two brothers obtaining scholar-
ships at the same time. I am afraid it is rather
ill-natured, but I must confess it adds a little to my
gratification, that we have beaten several public
school men, who are apt to look down with contempt
on those who have been educated at home. The
number of scholars elected (exclusive of the Westminster scholars, who do not pass the same examination with the rest, but are elected before they
come up to the University is sixteen, eight of the
third year and eight of our year. We were all
marched into chapel on Friday morning, and took
the oaths before the eight seniors, and yesterday we
took our places in hall at the scholars' table,
where we seem likely to fare better than we have
done hitherto.

In my hurry in leaving Barton on Thursday, I
forgot to leave my dried plants where Emily could
get at them.

1831   I suppose you will receive this letter just as you are in the fever of expectation of the grand debate; I shall look anxiously for the news on Tuesday.

Believe me,
Ever your very affectionate son,
C. J. T. BUNBURY.

———

During the vacation Charles and his brother Edward went to stay at Linton on the north coast of Devonshire to read.

———

Linton
July 1st, 1831.

My DEAR FATHER,

We had a very disagreeable passage hither from Bristol on Wednesday, the weather being so rough that we were both extremely sea-sick, and it was not without a good deal of difficulty that we could land. On arriving, we found to our great satisfaction that Burgoyne had joined the party. We are now settled in comfortable lodgings, in the same house with him.

Yesterday I did nothing but ramble and explore, but to-day I have begun reading, and hope to get on pretty well, though the delay of our books is a great inconvenience.

Our post town is Barnstaple, at the convenient

distance of two-and-twenty miles. Any parcels that 1831
you may want to send us, had better be dispatched
by the coach to Barnstaple, as there is a carrier
from thence three times a week. We get very
tolerable tea and sugar here (that is, at Lynmouth),
as also stationery.

I am very much pleased with this place, which is
really beautiful, but in quite a different style from
Ilfracombe. Lynmouth is situated in a kind of cove,
at the mouth of the Lyn, with high hills, partly
wooded and partly rocky, on three sides of it ; and
from thence it is a good quarter of an hour's walk,
up a very steep hill, to this place ; and even this is
not at the top of the hill. In Italy, Linton would
be called the Borgo, and Lynmouth the Marina.

The Lyn is formed by two streams, which join
just before they reach the sea. I walked about two
miles up the course of one of them, the East Lyn,
and was delighted with my ramble. The stream is
rapid and very clear, winding and murmuring among
great stones and broken rocks, and the hills rise very
steeply from it ; those on the one bank being rocky
and bold, on the other, richly clothed with oak-wood.
The frequent windings and turns of the valley add
to its beauty. The famous valley of Rocks is wild
and singular, rather than beautiful ; in fact, in the
valley itself there is nothing remarkable, but the hills
which bound it are crested with curious tower-like
piles of rocks, somewhat resembling the Tors on
Dartmoor. Where the valley opens to the sea there
is a magnificent castellated crag, the foot of which is

1832 washed by the waves, and just beyond it a beautiful little cove. In general the fault of this place is that there are so few points where one can get down to the sea, or where there is any beach at all, consequently it is a bad place for bathing.

I hope you will write me a line as soon as you are at leisure and tell me how Emily and Cissy are.

Give my love to them, and

Believe me your very affectionate son,

C. J. F. BUNBURY.

———

In September, 1832, Charles Bunbury and his brother Edward visited Wales. He kept a journal of this tour, and gives a full account of the scenery, botany and geology of the different places which they visited. In the autumn he caught cold, and suffered much from a painful abscess in his right ear, which from being somewhat neglected, occasioned a deafness in that ear, from which he never recovered.

He was a good deal weakened by this cold, and his father dissuaded him from taking his degree, fearing the effect which the strain of work might have on his health. In some respects this was perhaps to be regretted, as it was believed he would have taken a first class degree.

His father was very anxious that he should enter political life early, and take an active part in the

affairs of the country, but before he did so, it was 1833
thought desirable that he should visit foreign countries; and from his fondness for botany, and from
his uncle Mr. Fox, being Minister at Rio de Janeiro,
it was settled that he should go there.

F. J. B.

———

Funchal,
Monday, June 17th, 1833.

My DEAR FATHER,

Here I am safe and sound in the capital of
Madeira, and a very strange-looking capital it is.
We landed here about eleven this morning, having
had on the whole a very favourable passage, though
with very little wind latterly, and to-morrow morning we are to re-embark. I suffered somewhat from
sea sickness during the first five days, and was
nearly a prisoner in my cabin, as I found the sickness could be pretty well kept off by lying still, and
though I was certainly not comfortable during that
time, yet I suffered less than I expected, and slept
well at night. It is curious how comparative comfort is to lie stowed away in a little ill-scented
pigeon-hole six feet by three (which they dignify by
the name of a cabin) is certainly not luxury, but it
is much better than sea-sickness. Latterly I have
felt no qualms, have recovered my appetite, and
have been able to admire the beauty of the nights,

F

1833 which have been lovely enough to make up for more inconvenience than I have suffered ; last night I stood a long time admiring the phosphorescence of ths sea, which showed itself very prettily, in frequent and bright but momentary flashes. After a ten days' voyage I should at any rate have been ready to admire Madeira, but it is really a very picturesque and beautiful island, all a mass of mountains, the sides of which, steep and rugged as they are, are very well cultivated wherever it is possible for anything to grow, and, as seen from the water, appear beautifully variegated with vineyards, corn-land, wood and rock. The tops of the mountains, wherever they are not hidden by the clouds, appear very rugged and savage, the shores are almost every-where high and precipitous, and the water of great depth even to the very foot of the cliffs. What I have seen of this island (for we coasted it for some hours this morning) is not unlike some parts of the Genoese coast, but the strange white forms and varied colours of the volcanic rocks render its appearance still more striking.

When we landed here all was new and strange and amusing—the narrow, crooked, pebble-paved streets, the old shabby looking houses, with their lattices and worm-eaten balconies, the wild uncouth look of the half-naked, mahogany coloured inhabitants, the vast leaves of the banana waving over the garden walls—at every instant there was something to attract my attention. But I must do the aforesaid mahogany ragamuffins the justice to say, that I never met with a more civil set of people, 1833 not a peasant but pulls off his queer little cupola-shaped cap when he meets you. Soon after landing, I and one of my fellow-passengers set off on horseback up the mountains, and we had such a ride as I never had before, up and down paved lanes that were almost perpendicular, and across ravines by paths where I should never have imagined that any horse could possibly tread. But I did see the most magnificent scenery, the finest mixture of the rich and the picturesque that can be imagined, and feasted my eyes with the sight of fuchsia hedges, and gardens full of bananas and Indian figs. In short, I am delighted with Madeira and have laid in a fresh stock of spirits to carry me on to Rio. I must not in gratitude forget to mention that Mr. Gordon, a Madeira merchant, who was one of my fellow passengers hither, has asked me to dine with him to-day, given me a most excellent dinner, and moreover promised to forward this letter to England by the first opportunity, for there is no regular packet *from* Madeira.

Ever your most affectionate son,

C. J. F. Bunbury.

P.S.—Love to dear Emily and to Edward if he is still with you.

1833   Rio de Janeiro,*

                                    July 24th, 1833.

My Dear Father,

You may talk as you please of Italy and Switzerland, but I never saw anything at all equal in beauty
to the neighbourhood of Rio, and I should doubt
whether the world can produce anything to match
it. The city itself lies rather low, but on both sides
of it, as well as opposite, gentle hills covered with
gardens, and white houses, extend along the shores
of the bay, and behind them rise the most picturesque
mountains, ridge beyond ridge, and peak beyond
peak, in endless succession, some rounded, some
strangely jagged and serrated, some shooting up into
sharp naked peaks and spires of rock, and all clothed
for the greater part with thick forests. At the mouth
of the bay these mountains approach so near together
as to leave but a narrow entrance, guarded on one
side by a very remarkable mountain called the Sugar-
loaf, a sharp regularly-formed pyramid of rock,
almost without vegetation, rising boldly from the
sea to the height of 900 feet.

From this narrow mouth the bay spreads into a
vast sheet of water, like a great lake, enlivened with
innumerable vessels of all sizes, and studded with a
great number of wooded islands, and the view is
terminated by the lofty range of the Organ Mountains,
dimly seen in the blue distance. Imagine all this
with such a sky and such a colouring as Italy can

---

* I omit his journal in Brazil for want of space.

hardly equal, and with all the richness and magnifi- 1833
cence of tropical vegetation. I am afraid my
description will give you but a very faint idea of this
glorious scenery, but I shall never forget it, and if
the voyage had been much longer and more
unpleasant than it was, I should have thought
myself amply repaid by what I have seen since I
landed here. How I wish that you and Emily and
Edward were here to enjoy the scenery with me.
But Emily would not like to stay long, for the town
is very noisy, and the heat is great; even now it is
what we should call in England very hot summer
weather, and this is about the coolest time of
year. The mornings and evenings however are
pleasantly cool, and I find the weather agree
very well with me, nor does it prevent my taking a
good deal of exercise; two different days I have
been on foot for five hours together, botanizing on
the wooded hills near the city I should tire you
with my raptures if I were to enter on the subject
of botany, or attempt to describe the extraordinary
richness and variety of the vegetation in the woods
about here, the strange and infinitely varied forms
and foliage of the trees, or the singular appearance
of the climbing plants which rise to the tops of the
highest trees, looking like great cables, and of the
epiphytes which, hanging from branches, adorn the
old trunks, and mantle the moist shady rocks. All
the pictures that I have so often drawn in my imagination fall far short of the reality; my expectations
are more than fulfilled, though I have not been more

1833 than a few miles from the city. It is true that at this season there are comparatively few plants in flower, but as everything I meet with here is new to me, I have plenty of occupation, particularly among the ferns, of which there is a profusion in the woods. I have been rather surprised with meeting with a considerable variety of mosses and lichens, which one hardly expects to see in a tropical country.

My dear Father, I cannot express how thankful I am to you for indulging me with a visit to this delightful country, the recollections of which will be an enjoyment to me as long as I live.

July 27th. You will perhaps be surprised that I have written thus far without mentioning my uncle,* but it will more surprise you to hear that he has not yet arrived. He is not missing however, for Mr. Ouseley heard from him a few days ago from Porto Alegre, and he is daily expected here—but so indeed he has been for some time past. I have met with the greatest possible kindness and attention from Patrick Blake's† friend Mr. Young, and from his son and daughter-in-law, who are very agreeable people.

If this letter finds you in England, and you happen to see Patrick, pray remember me to him, and say that I thank him most cordially for his letter of introduction, which has been of the greatest service to me. I was surprised the other day by a visit from the English Admiral here, Sir Michael Seymour,

---

* Henry Fox, Minister at Rio de Janeiro.
† His cousin Admiral Blake.

and now he has asked me to dine on board the 1833 *Spartiate* tomorrow.

Hanmer's *favourite* brig, the Snake, is lately arrived here. I am at present comfortably lodged in a good hotel kept by a Scotch woman, which has no other disadvantage than that of being in rather a noisy and dirty street, and quite in the heart of the city, so that one has some way to go in order to escape into the country. When I say the street is *rather* noisy and dirty, I speak comparatively, for the whole of Rio is both the one and the other, in a great degree. You have no idea what a horrid noise the blacks make, particularly when they carry burdens through the streets.

I will take this opportunity of thanking Emily for her Shakespeare, which was nectar and ambrosia to me on the voyage. I long for the arrival of the next packet, to know how dear Sarah is going on, and how you and Emily are yourselves.

I had almost forgotten to mention that the disturbances in Minas Geraes which we heard of just before I sailed have been quieted, and there seems now to be no obstacle to travelling there. At Pernambuco there was a row some little time ago, and indeed one can expect nothing else in a country so vast as this, and where the government is so feeble. I visited the Chamber of Deputies the other day; it is a handsome room, and an orderly gentleman-like looking assembly, but unfortunately the wit and wisdom of the speakers was thrown away upon me, as I have learned but little Portuguese.

1833 *July* 30*th*. I heard yesterday that my uncle was arrived at S. Paulo, so he really seems to be making progress hitherwards. My dinner with the Admiral on Sunday was pleasant enough; he seems a very good-natured, mild-mannered, gentleman-like man. On Thursday next I dine with Mr. Ousely, whom I have not yet seen, but who appears to be very much liked here. Among many other kindnesses, Mr. John Young has got me admitted to the English reading-room, where I have the opportunity of consulting many books that will be useful to me. In short, I am as comfortable as it is possible to be, (for even the mosquitoes have not yet annoyed me much), and in spite of the bustle I was in latterly in London, I do not find I have left anything material behind me, nor do I want anything from England except some of my old friends, who I am afraid will hardly be persuaded to come. Giovanni is a most excellent servant. Pray give my best love to Emily, Edward, Cecilia, &c. I will write to Edward by the next packet.

Ever your very affectionate son,

C. J. F. BUNBURY.

———

Rio de Janeiro,
August 22nd, 1833.

MY DEAR FATHER,

His Excellency is come at last; he arrived last Saturday, and was, as you may suppose, a good deal surprised at finding me here. He came in an

American merchantman that he had hired, and had 1833 touched at a number of places in his way, staying some time at Porto Alegre, St. Catherine's, and St. Sebastian, which accounts for his delay. He is become a great botanist, and has made a fine collection of plants at the different places where he stopped, many of which are probably new, and he has promised me duplicates of them all. His account of Buenos Ayres has determined me not to go thither, he says it is the most horrible place in the world, without one single attraction of any sort or kind, and as for scenery, he says, you have on one side a river that looks like a muddy sea, on the other a goose-common 1200 miles wide, in comparison with which Newmarket Heath is mountainous and picturesque. The journey across the Pampas is become impracticable, or at least extremely dangerous, by reason of the Indians.

At Buenos Ayres he made acquaintance with Bonpland, who is grown very un-European, and so fond of Paraguay that he is going back there. I think I told you in my former letter, that I had not yet seen Mr. Ouseley, he has since shewn me a great deal of civility and kindness, and I like him very much; he is a very gentleman-like man, with a singularly mild, courteous, pleasing manner, and a great deal of information and good sense. His political opinions are very liberal, and in particular I was glad to hear him say that after having lived several years in slave countries, he is more than ever convinced of the evils of the slavery system, and the necessity of its

1833 speedy abolition. Mrs. Ouseley who is an American, is very lady-like, pleasing and pretty.

August 30*th.* I returned on Wednesday from a visit to Mr. March's place in the Organ Mountains. My excursion was very interesting and initiated me a little into the mode of travelling in this country. I had the opportunity of seeing a Brazilian forest in all its glory, and a wonderful scene it is, well worth coming all this distance to see, the amazing height of the trees, the endless variety of their forms and foliage, the strange appearance of the innumerable epiphytes, that load their trunks and branches, the huge snake-like climbers, the excessive thickness of the underwood, the beauty of the bamboos, fern trees and palms, are things of which one cannot form an adequate idea without having seen them. Mr. March's farm is situated on a kind of table-land, or as it were on the first floor of the mountains, amidst beautiful green pastures, with magnificent peaks towering above. The elevation above the sea is about 3,100 feet, and the climate is so much cooler than that of Rio, that all kinds of European fruits and vegetables are cultivated with success, and the thermometer sometimes falls nearly to the freezing point. Our return was rather fatiguing. On Tuesday I rode 30 miles in the rain, and part of the way was down an excessively steep rocky path, little better than a water-course, which would be almost impracticable for anything but a mule.

Up to the beginning of this week we had had a long continuance of very hot weather—unusually hot

for this time of year, as I understand—the ther- 1833
mometer every day at 82 or 83 degrees in the shade ;
now it is a good deal cooler, and we seem likely to
have much rain. I am as well as ever I was in my
life, and I do not find that the climate makes me at
all languid or disinclined to take exercise.

I had almost forgotten to tell you of a negro merry-
making that I saw at Mr. March's. It was by far the
strangest scene I ever witnessed ; there was a good
number of blacks, men and women, dancing by fire-
light in a kind of outhouse, and what with the
irregular red gleams of the fire on their uncouth
figures, the darkness of the rest of the buildings, the
extraordinary contortions of the performers, the
strange wild monotonous chant and violent clapping
of hands, it gave me more the idea of a witch's
sabbath than anything I ever saw. I often wish that
I had your talent for drawing, to sketch some of the
innumerable strange figures that one sees here.

September 2nd. The packet arrived at last, yesterday, and brought me your kind letter, which was
very welcome indeed. It grieves me to hear that
you have yourself been so unwell, I hope Italy will
set you quite right again. Though I am here in the
midst of all the magnificence of tropical vegetation,
I cannot read of the flowers and fruits of Barton
garden, without longing to be among them.

Pray give my best love to dear Emily and George.*

Ever your most affectionate,

C. J. F. BUNBURY.

* General Sir George Napier, K.C.B.

1833

Rio de Janeiro,

Thursday, October 3rd, 1833.

My Dear Father,

A thousand thanks for your letter. I am much shocked and grieved at the account which both your letter, and one that I received at the same time from my poor dear aunt, give of Fanny.* It is melancholy to think of particularly as regards my poor aunt, who has already had so much sorrow and anxiety to bear.

You do not mention your own health, therefore I hope and conclude it is better than when you wrote before. Respecting myself I have little to tell you I am very well in health, but subject now and then to painful twinges of *home-sickness*, which I must struggle against as well as I can, for it would be useless to return to England while you are abroad, and besides I cannot in decency return, without seeing something of the interior of Brazil. It has been awfully hot for some time past, and at the same time very thick and hazy, so that we have been almost deprived of the enjoyment of the scenery. The rainy season will probably soon be setting in, and by all accounts it is useless attempting to travel while that lasts, for one can neither observe nor preserve anything—so I shall hardly be able to set out on my travels before March or April. I am a good deal dismayed at hearing you talk of remaining so long abroad. What shall I do if I return to England

* Sir William Napier's eldest daughter.

before you ? for I have no intention of remaining in 1833 this part of the world anything like three years.

Nothing particular of a political nature, has happened here since I last wrote, but about once every fortnight or three weeks there are rumours of an expected tumult, and the national guards are called out, and a proclamation issued, and the streets patrolled, and a few poor snobs taken up, and a certain degree of excitement produced, and then all is quiet again. It is just possible that *the wolf* may come at last, but I very much doubt it. Indeed I cannot conceive that any revolution can happen here of a nature to be at all alarming or dangerous to strangers, unless it be an insurrection of the slaves. Talking of slaves I must tell you what a Brazilian gentleman farmer at St. Catherine's said to my uncle. After bitterly lamenting and inveighing against the suppression of the slave-trade, he said that as he understood that both England and France had been revolutionized and were now under liberal governments, he hoped they would restore the freedom of the said slave-trade, and remove the obstacles which had been put in the way of it. Rather a new idea of *free trade*. Though I never took any great interest in the Portuguese contest, I am glad it has turned out in favour of Don Pedro, but I think I am more pleased at the decisive blow having been struck by a Napier, than at anything else in the business. I am sorry the motion for Triennial Parliaments was thrown out, though perhaps one had not much right to expect it would be carried at present ; it will be

1833 before very long I have no doubt. What an aristocratic speech Lord John Russell made on the occasion! one can hardly recognize the reformer in it. From the second reading of the Irish Church Bill being carried in the House of Lords, I infer that the Tories despair of being able to make up a Ministry if they should throw Lord Grey out.

I am very glad that Henry is promoted, and going to so pleasant a place as the Cape is said to be, and I cannot help flattering myself that he may perhaps touch here in his way out.

October 12th. The weather grows more and more disagreeable—a constant fog, sometimes so thick as to remind one of an English November, bnt intensely hot, night as well as day; last Thursday the thermometer was at 80 at ten in the evening in a large room, with every contrivance for a thorough draught. If I return from this country without being entirely melted, I think I shall never wish for warm weather again. The mosquitoes too are very active just now, and do not contribute to make the heat more endurable. I caught a fine diamond beetle yesterday in the course of my walk. How little I thought years ago, when I used to admire the beauty of the diamond beetle in cabinets of natural history, that I should ever have an opportunity of seeing one alive. Snakes are nothing like so common here as I had expected to find them; I have seen but two in the three months I have been here, though I am continually rambling about the hills and in the

woods.   Give my best love to Emily and the rest 1833
of your party, and believe me,

Ever your very affectionate son,

C. J. F. BUNBURY.

Rio de Janeiro,
Friday, November 22nd.

MY DEAR FATHER,

I have taken my passage to Buenos Ayres by the
next packet.   I do not intend to stay there long,
but I shall probably find enough in the way of botany
to amuse me for a month or so, and at any rate it
will be pleasant as a variety, and as Buenos Ayres
is a place of some notoriety, it would be a pity to
return to England without seeing it.   You can have
no idea how disagreeable the weather has been since
I last wrote, till this last day or two ; a thick yellow
smoky-looking fog hiding all the landscape, and
making the heat ten times more stifling and oppressive, the sun appearing through it no bigger than the
moon, and of that lurid red colour which he sometimes puts on in London.   I have often thought of
the Ancient Mariner, " All in a hot and copper sky,"
etc.   If I were to judge from my own experience, I
should say that the climate of Rio was abominable
beyond all abomination, but I understand this is a
very extraordinary season.   I often ride to the
Botanic Garden, which is about seven miles off, in a
beautiful situation ; it is not a rich collection but

1833 there are some very fine Eastern plants, and it is interesting to see the Bamboo, the Cinnamon, the Bread-fruit, and the Jack-tree flourishing luxuriantly in the open air. Another favourite ride of mine, is along a broad sandy beach beyond the mouth of the harbour, with noble rocky mountains on one hand, and the open sea on the other, with a fine surf breaking on the shore. It is a delightfully wild and lonely place, frequented by nothing but sea gulls. I have not observed here very much that is interesting in the way of geology; the only rocks are granite and gneiss, which seem to pass into each other, but the granite is the more abundant of the two, and is in some places of a very fine quality. The gneiss has a very pretty appearance from the great quantity of small red garnets imbedded in it. I have met with no other minerals, not even crystalized quartz. It is extraordinary that with an inexhaustible supply of such fine materials for road making close at hand, the roads about Rio should be so grievously bad as they are —and yet I am wrong, it is *not* extraordinary, for these people would rather submit to any inconvenience than take the trouble to remedy it. The granite for building is brought into the city in rude bullock-carts with very clumsy wheels, and as the Brazilians think it improper to grease these wheels they creak and groan, and scream in the most diabolical way. I do not know whether I may not have mentioned this before, but it is a nuisance that continually forces itself upon one's attention.

November 23rd. It has been raining for the last

two days without intermission, and as hard as it can 1833 pour; the whole atmosphere seems as if it were converted into water, and the streets of the town have become rivers.

Pray give my best love to all your party. I suppose Edward is with you by this time and hope he has got quite well. My next letter will be dated (barring accidents) from Buenos Ayres.

Ever your very affectionate son,
C. J. F. BUNBURY.

———

Rio de Janeiro,
Friday, November 1st, 1833.

MY DEAR EMILY,

I am very much obliged to you indeed for your very amusing letter, which I will lose no time in answering; I have several times before thought of writing to you, but I was deterred by the bad accounts I received of dear Sarah,* thinking that my letters would not be welcome while you must be so anxious about her, I cannot tell you how happy I am to hear that she is so much better. Your account of the books you have been reading interested me very much but at the same time was rather tantalizing, for you may well suppose I have no opportunity of meeting with new books here. I am the more likely however to become well acquainted with those that I brought with me and I will take this opportunity

* The eldest daughter of Sir George Napier.

1833 of thanking you heartily for your Shakespeare, which is really a treasure to me, and has charmed away many a tedious hour, both on the voyage and since I arrived here. The more I study him, the more I find to admire. One thing that particularly strikes me in him is that he does not, like most writers, finish carefully only one or two principal characters, and leave the rest undistinguishable, but (in his best plays at least) the subordinate characters have as much life and reality and individual existence as the principal ones. For instance, in Henry the Fourth, Justice Shallow and his man Davy are as perfect in their way as Prince Henry and Hotspur. I do not much admire Johnson's Preface, he does not seem to me to appre iate or point out properly the real distinguishing merits of Shakespeare. Though I cannot boast much of my proficiency in astronomy, I certainly was rather struck with the different appearance of this hemisphere, particularly in that part of the voyage when we could see a portion of each; but we northerners have considerably the advantage —the southern hemisphere is much less brilliant than the other, and I was particularly disappointed in the Southern Cross, which I had imagined to be something very splendid. I do not know anything that impresses so strongly on my mind the feeling of the great distance which now separates me from my country and friends, as the reflection, which often occurs to me here when I am looking at the stars, that my friends in Europe cannot even see the same stars. Perhaps it will not be disagreeable to you if

I say a few words about botany, which is the chief 1833 thing I have to attend to here. Of all the strange plants of this country, those that have made the most impression on me are the tree-ferns and the bamboos, of which I had a very imperfect idea before ; even our beautiful Osmunda regalis gives no idea of the tropical tree-ferns, for their mode of growth is quite different ; they rise to the height of fifteen or twenty feet, with a slender columnar stem like that of a palm-tree, crowned with a circle of immensely large and finely divided leaves of the most delicate pale green, dropping gracefully like ostrich plumes. Though less majestic than the palms, they are more delicate and beautiful. The variety of ferns hereabouts is very great ; I have already found above thirty species within a walk of the town ; and the Organ Mountains are still more rich in them. The Cecropia is one of the most common trees in these woods and a very strange looking tree it is, with its long, slender, naked branches placed in whorls, and each terminated by a single tuft of large palmated silvery leaves. Uncle Henry* has taken a house about three miles out of the town, beautifully situated on a hill, looking to the sea, across a plain which appears one great garden, chequered with grass-fields and coffee plantations, with groves and clumps of orange and mango trees, and patches of bananas and tall solitary palm-trees here and there ; the Sugar-loaf mountain full in view, and a

* His Excellency Henry Stephen Fox

1833 picturesque thickly-wooded mountain (called the
Corcovado) behind. By the way, I surprised all my
friends here by walking to the top of this same
Corcovado—not that there is any difficulty in it,
but the residents at Rio are not accustomed to such
long walks. I ride out pretty often. I am a little
surprised, I confess, that neither you nor my father
mention a word about politics in your letters ; I
should have been curious to know what he thinks of
the state of affairs. It is a pleasant instance of the
perversity of human nature that I feel more interested in English politics *now*, than I did when
it was in my power to have taken a part in them.
The politics of this place are very uninteresting.
I am sorry to hear that the gale has done so much
mischief to the trees at Barton, I hope the
Magnolias have not suffered. That gale seems to
have been terribly destructive on the English coasts.
Pray give my best love to dear Sarah and Cecilia,
and believe me,

Yours very affectionately,

C. J. F. BUNBURY.

Buenos Ayres,
December 31st, 1833.

My Dear Emily,

Shakespeare has given, in a few words, so complete a picture of Buenos Ayres as will save me a
world of description :—" Barren, barren ; beggars
all, beggars all ; marry, good air"—that is to say, where

it is not polluted by the smell of hides and horns, and 1834 all the refuse of the slaughter-house. The town is regular, well built, and handsome enough, but tiresome from its uniformity; the country a dreary flat; the people, handsome savages. I have been here just a fortnight, and am already tired of the place, but as I am here, I may as well stay long enough to make a tolerable collection of the plants of this neighbourhood, which I believe are not much known. I have already noticed here several European and even English plants, but whether they are real natives of the country, or merely naturalized I cannot undertake to say; at any rate, some of them are very plentiful here, particularly the common Fennel, which grows in vast abundance in the hedges, among the American Aloes. The vegetation altogether has a very European character, and one strikingly different from that of Rio Janeiro; almost all the native plants here are herbaceous, whereas *there* the great mass of the vegetation consists of trees and shrubs. Almost the only elevated objects in the neighbourhood of this town are the American Aloes, of which most of the hedges are made, and the long straight rows of these tall, stately, flowering stems (for they are now in full blossom) have a most singular appearance. I had fancied that the Pampas came quite up to the town of Buenos Ayres, but instead of that, the country for four or five miles inland from it, is all enclosed, and yet has nothing rich or cheerful in its aspect, being overrun with great thistles, and other coarse, unsightly weeds.

1834 The river is so wide that one cannot see across it,
and it might be taken for a sea, but it is of the most
filthy, muddy colour imaginable, and when angry
(which it very often is), each wave is crested with
tawny foam, like Scott's Border river. The soil all
about the town is a stiff clay, and a day of rain is
sufficient to make the roads almost impassable.
However, as it is but fair to shew you both sides of
the picture, this place has one or two advantages
over Rio Janeiro; and firstly, a more agreeable
climate. The weather since I came has been very
pleasant, sometimes very hot in the middle of the
day (yet nothing like the heat of Rio), but refresh-
ingly cool in the evening and night. The changes
of temperature indeed are frequent and rather sudden, so that the place does not suit all constitutions.
Secondly (which it was very ungallant of me not to
put first), the ladies are much handsomer; and thirdly,
the town itself is neater, cleaner, better built, better
paved, and infinitely less noisy. The houses are
almost all alike, low, flat-roofed, and solidly built,
enclosing one or more courts, and with iron bars
before their windows.

January 1st, 1834. A happy new year to you,
dear Emily, and many more such. I am very anxious
for letters, as I had none by the last packet, and I
hope to hear that dear Sarah is getting well in the
mild air of Pisa. So I hear the Spaniards are
beginning to cut one another's throats: it is an
amusement their Argentine descendants are very
fond of, which does not contribute to make a resi-

dence here more agreeable ; there has been a sort of 1834 revolution very lately and though all is settled for the present, the storm has left a kind of *swell* behind it. I have been strongly advised by Mr Gore* not to venture far from the town. Mr. Gore, as you probably know, is *Chargé d'Affaires* here, and a good-natured, agreeable man, but seemingly very much out of his element in South America ; he abhors this place and the people thereof, and says that all the wealth of South America would not induce him to remain here another year. In fact I have found very little difference of opinion among the English gentle-men whom I have heard speak of Buenos Ayres. I wish, however, that my father were here to make a sketch of a Gaucho on horseback, wrapped in his ample, many-coloured poncho, with his coal-black hair falling down on his shoulders, and his bare feet resting in silver-plated stirrups. Do you recollect a little rubbishy-looking plant, by name Coronopus didyma, that grows under walls and by way-sides at Dawlish ? That little vagrant I have gathered in Madeira, at Rio Janeiro, and here ; it seems to be quite a citizen of the world. Bonpland is gone back, not exactly to Paraguay, but to the Missions adjoining to that country, and it is said he intends to remain there five years, to complete his collections and observations. It may perhaps be doubted whether he will ever make up his mind to return to Europe. This place is rich in pretty bulbs,

* Afterwards Earl of Arran.

1834 Amaryllises, Sisyrinchiums, Morœas, &c, which give a very gay appearance to the turf by the river-side. Now perhaps you do not know what a Morœa is, but you know what a Marica is, and they are so much alike that I daresay you would pronounce the distinction, to be without a difference; moreover, for your further edification, I have taken the portrait of this Argentine Morœa, which is a very pretty, odd-looking, tawny-coloured flower.

Pray give my best love to Sarah,* and to my dear Lady Disdain*.

Ever yours most affectionately,

C. J. F. BUNBURY.

———

January 14*th*. I have just received a grand heap of letters, and am quite delighted to hear that dear Sarah is so completely recovered. What a relief and happiness it must be to you as well as to George! I did not like Pisa much when I was there, but I shall always think well of it in future. So Henry† and Caroline† are gone abroad again. I am not much surprised at it, for she always seemed to have a hankering after Italy.

---

* Sarah and Cecilia Napier.
† Captain and Mrs. Henry Napier.

1834

MY DEAR FATHER,

Very many thanks for your kind and interesting letter from Paris, which I have just received. I suppose Emily will let you see the enclosed letter to her, which contains all that I have to tell about this dull place. Almost as soon as I arrived here, I made enquiries about crossing the Pampas to Mendoza in a carriage, but I found that would be too expensive, and I have no inclination for a gallop of 900 miles. Then I thought of going some way up the Parana in a vessel, but on a careful calculation, it turned out that that plan also would cost rather more than I can afford, so I must content myself with Buenos Ayres and Monte Video. Though exceedingly dull in other respects, this is not a bad place for botanizing. Uncle Henry writes me word (by-the-bye is not that a wonder?) that the heat at Rio is greater than has been known for forty years past, the thermometer at 90 degrees in the coolest rooms nearly all day long. This had begun before I came away, the day I embarked was the very hottest I ever felt. The weather here is very changeable, with frequent thunder storms. I have now been six months in South America and have literally not had an hour's sickness, so that I really have great reason to be thankful. Six months hence I may probably be thinking of returning home, but my plans are still unsettled. I had a very tedious passage hither

1834 (eighteen days) with a pleasant alternation of calms
and squalls, so that if the commander of the packet
had not fortunately happened to be a very gentlemanlike and pleasant as well as a most obliging man,
I should have been but ill off. I have received by
this packet a long and exceedingly interesting letter
from Edward, and a melancholy one from my poor
dear aunt, who seems to have almost given up all
hope of Fanny's recovery. Most probably by the
time I write this it is all over. I am very glad that
the peace is likely to be kept in Europe, both for
private and public reasons. There was a sort of
petty revolution here just before I came; an army
of 6,000 or 7,000 Gauchos blockaded the town, and
the Government, not being able to muster more than
a few hundred men to oppose them, sent to General
Rosas, who is fighting the Indians on the borders of
Patagonia, to come and chastise the insurgents; his
answer was, that he thought the insurgents very
much in the right, and if he came at all, he should
fight for them, not against them. The Government
therefore thought it prudent to resign, and a new
one was set up according to the taste of the Gauchos.
It is believed that Rosas, when he has sufficiently
licked the Indians, will come here and take the power
into his own hands.

Ever your affectionate son,

C. J. F. BUNBURY.

Buenos Ayres,
January 16th, 1834.

MY DEAR UNCLE,

1834

Many thanks for your letter. You will have heard before this from Captain Coghlan, that I did not stop at all at Monte Video, but came on to this place, where I have been ever since. I must acknowledge the fidelity of your description of Buenos Ayres: it is a dull and detestable place, but I do not find it at all bad in the way of botany, and the climate is certainly pleasanter than that of Rio. Mr. Gore seems to hate the place full as much as you did. I had a great mind to hire a large boat and go up to San Nicolas, on the Parana, which is said to be a pleasant place, but it turned out that the expense would be more than I could afford. I am much struck with the European aspect of the vegetation here, so different from that about Rio, in fact I have observed many species identically the same that we have in England, and those so abundant that I cannot but suppose them to be real natives of the country. The turf by the river-side is at present very gay with a variety of pretty flowers, among others a minute Lobelia, a white Amaryllis (or Sternbergia of some authors), the Sisyrinchium Vermudianum, and a very pretty tawny yellow Morœa (Morœa Herberti I believe) which I think I saw in your collection under the name of Ferraria. The genera Morœa and Ferraria are indeed very nearly related. I have collected altogether, about forty species, of which a large proportion are Solaneæ, Amaranthaceæ and Grasses; a good many of

1834 them are undescribed, at least in "Sprengel's Systema," which is the newest general work. My walks have been a good deal limited by the frequent thunder-storms, which make the roads excessively miry.

I have a melancholy letter from my poor aunt, written from Penzance, whither she has carried Fanny for the chance that the mild air might do her good, but she seems to have very nearly lost all hope. She desires her kind love to you. The accounts from my father and Edward are very satisfactory: the latter had been making a tour in Hungary and was much pleased with it. Sarah Napier was completely recovered, thanks to the climate of Pisa. Henry was to sail in the course of this month for *New South Wales*, not as a convict, but as a Lieutenant in the 21st Regiment. Henry and Caroline Napier had let their place in Hertford-shire, and gone back to Italy. I shall be as usual guided chiefly by circumstances, but my present plan is to go before long to Monte Video, stay there till Captain Coghlan arrives, and return with him to Rio, as I suppose the worst of the heat will be over by that time, and at any rate the bracing effect of this climate will have made me more able to bear it. I saw a review of the troops this morning in the Cathedral square, and such extraordinary looking soldiers I never set eyes on before.

> Believe me,
> Your very affectionate nephew,
> C. J. F. BUNBURY.

Buenos Ayres, 1834
February 17th, 1834.

My Dear Father,

I expected to have been at Monte Video some time ago, but have been detained at this vile place by a succession of vexatious casualties. In the first place I was wind-bound for a week; then just as I was going on board the packet, a rascally custom-house officer chose to search me, and seize all my money, so I had to apply to Mr. Gore for redress. My money was restored next day, but in the meantime the packet had sailed, and since that time I have been dawdling on in expectation of the English packet, which ought to have been here some days ago. These petty annoyances are all I have to tell you of, as I have seen nothing worth mentioning since I wrote last, except the diversions of the Carnival, which are quite worthy of the Buenos Ayreans; they consist in throwing bucketsful of water from the balconies, and tops of the houses, upon passengers in the street, and pelting them with eggshells filled with water. I rejoice at the prospect of returning to Rio Janeiro, which will appear ten times more delightful after this abominable dungeon, and from thence I shall set out on my travels as soon as I can.

March 17th. I am just returned to Rio de Janeiro, I left Buenos Ayres on the 27th of last month, and had a very fair passage, stopping two days at Monte Video, which is just like Buenos Ayres on a small

1834 scale.   The country about here has been much
refreshed by the rain, and looks more beautiful than
ever.   The French ambassador at this court, the
Count de St. Priest, (who was my fellow passenger
in the Rinaldo) is gone to Buenos Ayres with his
secretary and his cook; one of the newspapers of
that place says that he is not gone on any diplomatic
mission, " but solely from a natural curiosity to see
a city which, for the magnificence of its buildings,
and the high civilization of its inhabitants, may
properly be called the Paris of South America." Poor
Paris !

My uncle is looking a good deal the worse for the
heat of the summer, though he says he has been
able to keep himself tolerably well by not stirring
out till near dark, and eating scarcely anything.   It
is now a good deal cooler than it has been, though
the sun is still fearfully powerful in the middle of
the day, so that one is a close prisoner from nine or
ten in the morning till four in the afternoon.   I am
excessively obliged to you for Horace Walpole's
Correspondence, which I found here on my return
from Buenos Ayres: how full of wit it is, and how
curious and interesting are the political details ! I
have not yet made any use of the gun you sent me,
it is likely to be useful when I go up the country, but
there is nothing here to shoot.   I believe indeed I
might meet with some snipes, if I chose to go forty
miles for them, and then wade up to my middle in
stagnant water under the full blaze of a tropical sun ;
but I am not yet tired of my life, and when I am, I

shall choose some easier and less troublesome kind 1834
of death.

March 23*rd*. The February packet has just
arrived—but *not* the January one—and I have to
thank you very sincerely for your kind letter of
January 16th and for your great kindness to me about
money. I am quite ashamed of the querulous letter
I wrote you in November, indeed I have great cause
to be thankful instead of complaining, having enjoyed
such uninterrupted good health, and met with so
much kindness from a number of people who were
before strangers to me. My design is to set off
about a month hence, when the cool season will be
fairly set in, and I shall have four or five months of
good travelling weather before me. I cannot say
I have any great desire to visit the United States;—
my curiosity about them has been pretty well
satisfied by what I have heard and read—but I will
certainly be guided by your wishes, and I fancy
there will be no difficulty whatever in obtaining a
passage thither from this port. I have received by
the same packet a very agreeable letter from Edward,
for which pray thank him in my name, but I am
very sorry indeed to hear that he continues so
unwell, and is paying so dear for his honours. As I
wrote to him from Buenos Ayres, and have at present
little or nothing to say, I must content myself with
sending him my best love, and will write to him before
setting out for the interior. I have also a letter from
Bessie,* with a few lines from my poor dear Aunt,

* Afterwards Countess of Arran.

1833 and I am happy to find that her health and spirits seem to have in some degree recovered from the sad shock of poor Fanny's death.

One of my fellow passengers from Monte Video hither was a very intelligent Brazilian gentleman, from whom I picked up a good deal of information. I must tell you one thing he said, which struck me much— that he was convinced that even in this climate a white man might, by working early in the morning, and late in the evening, do as much work in six hours as a black could do in twelve.

Pray give my best love to Emily and the rest of your party, and believe me,

Ever your very affectionate son,
C. J. F. BUNBURY.

P.S. *March 27th*. The January packet is at last arrived, and has brought me another delightful parcel of letters, how good it is of all of you to write to me so often! I am particularly obliged to Emily for her account of De Candolle, which interested me very much. I am surprised at Edward's liking Pisa so much—not at all surprised at your dis-liking it, from what I remember of the place.

April 7th. I have been ill, though not severely, with what the doctor calls a gastric fever, but am getting well again. It has been a great nuisance, having condemned me to complete idleness for nearly a fortnight, and this is not a place where one can be idle with comfort. My uncle has lent me a number of reviews and newspapers, which have served me in some measure to beguile the time, 1834 and in particular I have been much interested with the review of Crabbe's life. I am glad to hear that we are not to have a European war at present. I do not think I told you that there are rumours of a war between Buenos Ayres and Paraguay, a matter indeed of not much importance to the rest of the world, but it is supposed the Argentines will be glad of such an opportunity of finding employment for their rogues and vagabonds—that is to say for the greatest part of their population.

P.S.—If I go to the United States, I must send my collections home from hence.

———

Rio de Janeiro,
April 26th, 1834.

My Dear Emily,

I owe you at least a million of thanks for two most agreeable letters which I have received since I wrote to you from Buenos Ayres, the first of which gave me a very interesting account of M. De. Candolle. But to begin with the second—it is very odd that you should have fallen in with my friend the Cavaliere, who is certainly a very curious specimen of mankind. I laughed mightily at his portraits of my uncle and myself; mine you can judge of; you know that my utterance is not very rapid even in English, and you may conceive that when I try to speak French or Italian it is doubly slow. As for my uncle, whatever he may have been formerly, he

H

1834 is become a complete hermit ; he never pays visits,
scarcely ever dines out, never gives dinners or
parties of any kind, and lives—not exactly on roots
and water, but very temperately, not to say abstem-
iously. His hours indeed are of the oddest : he
breakfasts at one or two in the afternoon, and dines
after dark. He takes long solitary rides, and
botanizes indefatigably, and with great success ; I
go to see him every now and then and we talk over
our plants. He says he would not remain here for
ten millions a year, if he were obliged to mix in the
society of the place.

Now for botany : I have no doubt that your little
crocus with the two green spathes is *Ixia* Bulbocodium (or Trichonema Bulbocodium, to speak
according to the newest authorities), which, as I
remember, is common in Italy in the early spring,
and has quite the habit of a crocus. The crocus
minimus is certainly our Barton friend. The hills
about Rio have lately been in great beauty with the
blossoms of two magnificent shrubs or almost trees,
one of them a Cassia with large branches of gold-
coloured flowers, the other is of the Melastoma tribe,
and has much the look of a giant Rhododendron,
but the blossoms are of a far brighter and richer
purple than those of any Rhododendron I am acquainted with. A great many other plants are now
in flower, and I have made large additions to my
collection since I returned from Buenos Ayres. But
after all, though the vegetation of this country is on
so grand a scale, and its forms so various and

beautiful, I have seen nothing in the way of *flowers* 1834 at all equal to the enamelled pastures of the High Alps, which I hope you will see this summer. Alas! I did not see the eclipse on the 26th of December, for I was then buried in that vilest of all places, Buenos Ayres, and was not forewarned of it.

I have received a very gay, amusing letter from Henry* who was on the eve of sailing for New South Wales.

For lack of more interesting topics, I am driven to the weather, which has been cool and pleasant for the last fortnight, but more rainy than is convenient. I enjoy this place much more now than before I left it, as the cooler weather enables me to work and employ myself with some degree of comfort; but pleasant as Rio Janeiro is in some respects, it is not a good place for one whose besetting sin is indolence, for it is difficult to resist the enervating influence of the climate. I feel the effect of this on my mind more than on my body, for I take habitually more exercise than I used to do in England. The Count de St. Priest does not seem to be of my opinion as to the agreeableness of Rio; he is so disgusted with it that he has given up his appointment and is going back to France in a rage. He was one of the writers of the Cent et Un, and it is supposed he intends to write a book on the Hundred and One Miseries of Rio de Janeiro. I hear from England that your brother† Charles's book sells

* His brother Henry W. Bunbury.
† On Cephalonia

1834 extremely well, and is much admired. Uncle Henry
said it was sure to succeed, for people would always
be pleased with a clever attack upon a man in power,
however little they might care for the merits of the
question. By the way, the " great philosopher,"
with all his hermitship and his botany, is not a bit
less satirical, or more disposed to judge mildly of
people, than he was in Italy ; though as far as regards
myself, I have found him extremely good-natured,
and indeed kind. I have not seen Mr.* Whewell's
book, but have read a review of it in the *Edinburgh,*
not on the whole very favourable, though the
specimens quoted are good; but I would take your
word in the matter of books. I am curious to see
" David Crockett," having been much diverted by
some extracts from it in the newspapers. A book
that has delighted me highly (not a new one indeed,
but new to me) is Humboldt's Tableaux de la Nature,
which my uncle lent me. Before I went to the Rio
de la Plata I read through his Relation Historique,
which I had before only dipped into ; it is to my
thinking the most interesting book of travels in the
world.

As I am at the end of my matter, and do not.
wish to wiredraw it I must conclude with my love to
Sarah and Cecilia. Believe me yours very
affectionately,

Le petit Fox.†

---

* Bridgewater Treatise.
† " The Portugese Caviliere had called his uncle the ' great Fox ' and him the
" little Fox.' "

Rio de Janeiro,
May 3rd, 1834.   1834

My DEAR FATHER,

I wrote you word by the last packet that I had had an attack, though not a severe one, of a kind of fever which has been extremely prevalent here, and which left me rather weak for a time. I have however completely recovered my strength, and have been waiting for fine weather to set out for the Mines; for the last fortnight we have scarcely had twenty-four hours together without rain, and the roads are in a deplorable state, but the weather may be expected soon to settle, and I shall then lose as little time as possible in getting under weigh. In the meantime, this neighbourhood is very pleasant, in the intervals of the rain, and the vegetable world is in great glory. I found the other day a most superb plant of the Orchis tribe, the finest I think I ever saw, and as it will not dry, I have *taken it off*, as we say in Suffolk, though I am half afraid you will think my drawing exaggerated when you see it. Your favourable opinion of the situation of the Ministry, which Emily mentioned in her letter to me, was previous to the meeting of Parliament, otherwise I should have been a good deal surprised at it, seeing the divisions that have taken place—one majority of eight (gained too by the help of the Tories) another of four, and a *minority* of six! As for the Pension question, I wish they had been actually beaten upon it, but at any rate that division will serve as a good *flapper* to them. Where were

1834 our members for West Suffolk? they do not appear
to have voted at all on the question. I shall have
some curiosity to see the division on the repeal of
the Union, though I cannot imagine that O'Connell
will get many English members to vote with him.
It seems the Malt Tax is not to be taken off, in spite
of the allusion in the King's speech to Agricultural
distress. I am just come from seeing the opening
of the Chambers here, and hearing the Regency's
speech, which however was read so fast, and so in-
distinctly that I could not understand a word of it.
There were all the Ministers and Officers of the
Household in uniforms plastered over with gold, the
Judges in black robes, with red collars, and a good
many priests all in black. The galleries of both
Chambers are open to the public without any impediment. The honourable Members receive a
certain pay during the session, besides having their
travelling expenses paid, which last must be an
important matter, considering the distance from
which some of them come; the deputy from Matto
Grosso, for instance, has a six month's journey from
his home to the capital. You see that annual
Parliaments would not answer in this country.

May 12th. The packet has been detained, so that
I am able to send you some more positive informa-
mation respecting my journey. I have bought a
riding mule for myself, another for my servant, and
two baggage-mules, all warranted as particularly
good, and a horse for the guide, for 800 milreis,
about £133; I have hired a guide, who is well

recommended, and has travelled the road very often, 1834 and I am to give him 20 milreis (about £3 7s.) per month, which seems reasonable enough, as he is to take care of the mules. Moreover, I have got my canteen filled, and shall take it with me, as well as two portmanteaus and some bedding, but the bedstead and tent I must leave behind, as they would require an extra mule. I set off in three or four days from this, and intend to be back again in about four months, by which time the hot season will be drawing near. As I am in for it, it would be a shame to return to England without having seen the Mines, but really one might travel over half Europe for the sum which is here swallowed up in the mere preliminaries of a journey! However I understand that when I once get away from Rio, the expenses of actual travelling will be very small, and if I bring back my mules in good condition, I shall probably be able to sell them again for nearly as much as I give for them. It will take me, I understand, about three weeks to reach Villa Rica, which I shall most likely make my head-quarters, for some time, and I will write to you from thence, but as the post in this country is slow, do not be alarmed if there are gaps in my correspondence. I went the other night to a ball given by an English merchant here, and found it tolerably pleasant, though to be sure there was a most plentiful lack of beauty. The weather for some days past has been delightful and seems likely to continue so. With the exception of my fortnight of sickness, I have enjoyed this place very much,

1834 since I returned from Buenos Ayres, and am half sorry to leave it; the pleasantness.of this season makes some amends for the frying that I endured in October and November. My friends the Youngs continue to shew me all possible kindness, and there is an English clergyman lately come out, who is a valuable addition to the society of the place. Believe me

Ever your most affectionate son,

C. J. F. BUNBURY.

From his Excellency HENRY STEPHEN FOX *to* SIR HENRY BUNBURY, *his brother-in-law.*

MY DEAR BUNBURY,

I hope this will find you and Lady Bunbury safely arrived and well established in Switzerland. You may judge from the newspaper accounts of our proceedings here, that I have been pretty well employed lately; a sad mess there has been, and it is to be hoped that we may gain wisdom from experience. I need not add that I am more than ever tired of the whole concern.

Ever your affectionate.

Ouro Preto,
June 10th, 1834.

MY DEAR FATHER,

At last I have the satisfaction of dating a letter to you from the Imperial City of Ouro Preto, better

known in Europe by its old name of Villa Rica. I 1834
arrived here yesterday in eighteen days from Rio,
and so far from being knocked up by the journey,
never felt stronger or in better health; in fact I
found the fatigue and inconveniences of the journey
much less than I had been led to expect, and the
interest of it far more than enough to make up for
them. My day's journeys were short, not above
eighteen miles in any instance, and I was fortunate
in having a remarkably quiet well-trained mule, so
that my attention was not distracted from the
observation of the country. But to proceed more
regularly;—I left Rio on the 23rd of May, early in
the morning, went by water to the head of the bay
and slept that night on the Serra da Estrella. For
nine days I travelled through woods of indescribable
luxuriance and beauty, continually up and down
hills, without seeing anything that could be called a
village, much less a town; the houses and spots
of cultivated ground, which are few and far between,
lie deep in the valleys and are not seen till one
comes near to them. On the 11th day, I got out
of the forests, and entered upon a high undulating
open grassy country, a good deal like the Wiltshire
Downs, and the same day reached the town of
Barbacena, but there was nothing to detain me
there. From the 2nd of June to the 7th inclusive,
my route lay over the Downs, which are not
beautiful, but amazingly rich in curious and interesting plants, of a totally different character from those
of the forests. Nothing certainly can be more

1834 complete or striking than the change in the vegetation ; instead of giant trees covered with parasites,
and luxuriant juicy climbers, there are short wiry
grasses, slender stiff shrubs, at most three feet high
with small leaves and gay flowers, and pretty little
herbaceous plants like those of the Genoese hills.
Even the woods which grow here and there in the
hollows, are of a dry and dwarfish nature, and quite
unlike the virgin forests.

The 8th I crossed the Serra de Ouro Branco, and
came into a more mountainous region, but still open
and bare, and thus continued to this city.   The
geological nature of the country may be briefly
described : the wooded mountains towards the coast
are of gniess ; the soil of the downs or campos is a
red clay full of fragments of quartz and brown iron-
stone, without any solid rock near the surface, and
the mountains on this side of Ouro Branco consist
of a soft kind of mica-slate.   For more details, I
must refer you to my journal,* which I have kept
very regularly.   This city would delight Edward, it
is so picturesquely situated, part in a deep valley,
part scrambling up the sides of half-a-dozen steep
hills, with the houses standing as it were one atop of
another, and surrounded by a wilderness of huge
barren mountains.   But a great town is not exactly
the best place for my pursuits, therefore I shall not
stay here above a week at most, but go on to the
English mines.   As I shall be constantly moving

---

* This journal I have not printed for the sake of brevity. F.J.B.

about, I shall not get any letters from you till I 1834
return to Rio de Janeiro, which will probably be in
September, but I hope you are enjoying your
summer in Switzerland as much as I am my *winter*
in the interior of Brazil. Really if I had but had
an agreeable companion, this would have been one
of the pleasantest journeys I ever made. I have
slept sometimes in farm houses, sometimes in little
clay huts which admitted no light but by the door,
and the chinks of the roof, and once almost in the
open air, that is to say under a rancho, which is
merely a roof supported on posts without any walls.
What you would perhaps not expect is that I have
suffered more from the cold of the night and early
morning than from anything else. The heat has
seldom been inconvenient. Anyone who judged of
Brazil from the city of Rio, would suppose it to be
much farther advanced in civilization than it really
is; the people in the interior live (or get along, as
the Americans say) in a very rough primitive way,
and seem quite contented with the absolute
necessaries of life. A man who has a large farm
and numerous slaves, will live on mandiocca flour
and black beans, and sleep on a mat or an ox-hide.
I must not forget to mention that Giovanni* has been
quite invaluable to me on this journey, so careful,
active, and obliging; indeed, he and the guide have
taken all the trouble off my hands.

Among the pleasures of my journey has been that

---

* His Servant.

1834 of seeing a live Toucan, but I have not attempted to preserve any animals, plants and minerals being quite sufficient to occupy my time—indeed I have not even had leisure to draw any plants since I left Rio. Since I began this letter, I have rambled a good deal about the neighbourhood of the city and found, among other interesting things, a great many octahedral crystals of magnetic iron-ore.

Pray give my best love to Emily and Edward, for I suppose the Napier party are not with you now.

Believe me,
Your most affectionate son,
C. J. F. BUNBURY.

———

Gongo Soco,
June 28th, 1834.

MY DEAR FATHER,

I write to you from the house of Colonel Skerret the Commissioner of the English mines here, who has shewn me a degree of kindness and hospitality that I can never forget, and has insisted on my staying with him to the end of the month. I have been here five days already, and have no hesitation in saying that they are the five pleasantest days I have passed since I left England. Colonel Skerret and his family have the happy talent of making one feel quite at home, and at one's ease, and you may imagine how delightful it is to me to find myself in such a kind and charming family, and in a comfortable *English* house, after having been a month rambling through an uncivilized country without any 1834 society than that of my plants and books. I have made great friends with the Colonel's children who are very nice little girls indeed, and remind me of my cousins* at Freshford, though not so pretty. To confess the truth, I have been pretty considerably idle since I came here, yet I do not consider my time thrown away, for I have learnt much from Colonel Skerret's conversation, and from what I have seen here, especially with respect to the negroes, I am every day more and more confirmed in my opinion of the benefits of emancipation. In England I was an Emancipationist from feeling, but my reason is now convinced. Colonel Skerret has freed several of the blacks employed about the mine here, and he finds that they work more diligently and steadily than they did before, and are in every respect more useful and trustworthy; but indeed it is a pleasure to see the blacks here, even the slaves, they appear so comfortable and prosperous and contented, and so much attached to those who have the care of them,— such a contrast to the slaves of the Brazilians! I have not been down into the mine, which is of great depth, but I have seen numerous specimens of the gold in its native state, and all the processes by which it is separated and freed from its matrix. It occurs here disseminated in a substance called jacutinga, a mixture of micaceous iron-ore, and granular quartz, which is extremely common in this part of Brazil, and forms entire

---

* The children of Colonel William Napier.

1834 mountains; the gold is generally in small particles, and not so beautiful as in the specimens from Transylvania. In other places not far from Gongo Soco, it occurs in quartz rock, but the mines in the jacutinga are always the richest. I wrote to you last from Villa Rica, where I remained five days, and botanized with great success, indeed I have not hitherto seen any place so interesting in that respect, and I shall make rather a longer stay there on my return. From thence I went by Marianna and Inficionado to Cocaes, where there is a rich gold mine which has been lately purchased by an English company; there, as well as here, the gold occurs in micaceous iron. I went down into the mine, not by a shaft, but by a low narrow passage, in which I was obliged to walk almost bent double, and sometimes to creep on hands and knees, and at last all that I saw was a dark dirty hole, and two or three naked blacks, working by the light of a farthing candle. From Cocaes I came hither in one day. By-the-bye, I do not think you will find Gongo Soco in the map, but it is about 20 miles east of Sabara; which is, or ought to be set down in every map of Brazil. From my leaving Villa Rica, till my arrival here, I met with little to please or interest me, and was growing rather sick of Brazilian travelling, and still more of solitude, but the pleasures of Gongo Soco make up for everything. The mountains among which this village is situated are mostly covered with woods of great beauty, but I do not find them nearly so rich in curious plants (at this season at least) as the campos or open country; however I have collected here a good 1834 many ferns, and other cryptogamous plants, some of them very beautiful. It is curious that wherever the woods have been cleared in this country, there are certain plants that seem invariably to spring up, and in a few years overrun the land, to the exclusion of everything else. I forgot to tell you that while I was at Villa Rica, I went to the highest mountain there, the Itacolumi, which is between five and six thousand feet high, and a very pleasant walk I had, and found a great many strange plants. When I leave this oasis in the desert, I mean to return by way of Sabara to Villa Rica, go from thence to S. Joao d'El Rey, and perhaps to St. Paulo, and so back to Rio de Janeiro. At present I do not recollect that I have anything more to tell you, except that I continue quite well and strong. Pray give my love to Emily, Edward, &c., and believe me,

Your very affectionate son,

C. J. F. BUNBURY.

———

Villa Rica,
June 11th, 1834.

MY DEAR UNCLE,

Here I am safely arrived in the capital of Minas Geraes after a very pleasant and interesting journey of 18 days, of which I found the fatigue and inconveniences much less than I had expected, and indeed

1834 quite trifling. The only thing from which I have suffered at all has been cold, which one would hardly expect within the tropics. It is true that I should have been uncomfortable enough if I had not brought some bedding, for one seldom meets with anything else by way of a bed than a mat or ox-hide; laid upon a wooden frame; but in the matter of eating I have fared very well. The first night I slept near the top of the Serra da Estrella and for the next nine days travelled through the forests, over a succession of high steep hills, without seeing anything like a village; the woods are magnificent, but there is very little variety in the scenery. In this part of the journey I collected but few plants and found it very difficult to dry them. At length I got out of the forests and entered upon the high undulating campos, which are like our English downs and where I found a prodigious number of plants quite new to me, resembling in their general appearance those that you brought from Rio Grande. The most abundant tribes are Melastomaceœ and Compositœ, of which the variety seems endless; there are also two beautiful Gentianeæ (Lisianthus speciosus and pedunculatus) several Rubiaceœ Lythrariœ, &c. These downs consists of a red clay formation, containing loose pieces of quartz and ironstone, without any solid rock near the surface. The 8th of June I crossed the Serra de Oura Branco, which consists of soft silvery mica-slate and entered a region of high grassy mountains which continues to this place. I have been here two days and found

much to interest me, especially in the way of 1834
geology; the various modifications of mica-slate in
these mountains are curious, and I have found a
great many perfect though small crystals of magnetic
iron-ore. Nevertheless I do not expect to stay
here above three or four days longer. My days'
journeys have been so short that I have always had
some hours' time for botanizing after arriving at
the sleeping-place, and since I came out of the
forests I have found no more difficulty in drying
plants than when I was stationed at Rio. Many
of the plants of the Campos may be thoroughly
dried in two or three days. In the three weeks
since leaving Rio, I have dried between thirty and
forty species and am daily collecting new ones;
among the number is that orange Bignonia which is
cultivated in the gardens at Rio; it grows in the
low wooded hollows between the downs, and is
very ornamental. The Vellosias I have seen in
one place only, on the Serra de Ouro Branco, but
there in great plenty; they seem to be very local,
and are certainly very singular looking plants. I
hope your botanizing has been as successful this
month as you expected.

Villa Rica or Ouro Preto is a large town, very
irregularly built, not handsome, but very picturesque,
buried deep among high, barren, rugged mountains.
I am tolerably well lodged here, though not exactly
in European style. Barbacena seemed but a paltry
town, and Quelluz is a mere village. From Porto da

1834 Estrella to Barbacena I did not see a house with glass windows.

> Believe me,
> > Your very affectionate nephew,
> > > C. J. F. Bunbury.

———

Gongo Soco,
June 29th, 1834.

My Dear Uncle,

I arrived here on the 24th intending to stay a day or two, but Colonel Skerret and his family have treated me with such great kindness, and made me feel so much at home, that I have been induced to remain much longer than I designed at first. To say the truth, it is not a very good station for botany, being surrounded by wooded mountains of which the productions appear in general very similar to those of Rio, so that I have found very little that was new to me except some ferns and a beautiful Alstrœmeria. But I have been much interested in observing the machinery and works connected with the mine, and the different processes by which the gold is obtained pure. Both here and at Cocaes the gold occurs disseminated in iron-mica slate, or jacutinga, a rock which is very common in this part of Brazil, but of which I have found it difficult to obtain good specimens, it is so soft and crumbly. On the Serra de Luis Soares, between two and three miles from hence, gold occurs in the

micaceous granular quartz, or quartz-slate as it is 1834
called by Spix and Martius, but the mines in the
iron-mica are said to be always the richest. I came
hither from Villa Rica by way of Marianna,
Inficionado, Catas Altas and Cocaes, but found the
journey, from one cause or another, much less
pleasant than before, and had little success in
botanizing, so that I was growing quite weary of
travelling when I arrived here. Since I came to
Gongo Soco, I have heard that the convent on the
Serra de Caraca still subsists, and I shall probably
visit it when I leave this, though I ought to have
taken it in my way hither. The Serra itself I had
constantly in view for two days, and a most noble
mountain it is. While I was at Villa Rica, I
ascended the Itacolumi, and found many plants
that were new to me ; near the summit there was an
abundance of Vellosias, but all out of flower; they
are among the strangest-looking plants I ever saw,
and of a particularly stiff ungraceful appearance.
I shall make a longer stay at Villa Rica on my
return, having seen no other place equal to it for
variety of plants. I hear that there is a tolerable
road from S. Joao del Rey to S. Paulo, and that it
is a journey of only ten days, so I may probably
return that way. Just at present I am rather lame
from the bichos having got into one of my feet, but
otherwise I have had nothing the matter with me
since I left Rio.

Ever your affectionate nephew,

C. J. F. BUNBURY.

Gongo Soco,
July 9th, 1834.

MY DEAR UNCLE,

I hope you have received my two former letters,
one of the 11th of June from Ouro Preto, the other
of the 29th from this place.   I have staid here much
longer than I at first intended, for the cordial kindness of Colonel Skerrett and his family makes me
feel quite at home, and the time passes so pleasantly
and quickly that I can hardly persuade myself that
I have been here a fortnight.   The neighbourhood
proves on careful examination much more interesting
in a botanical view than it appeared at first; it is
particularly rich in ferns, of which I have already
collected about twenty species quite new to me, in
particular several beautifully delicate species of
Trichomanes and Hymenophyllum.   Arborescent
ferns are more abundant and of greater size than I
have seen them anywhere else, and supply the place
of palm-trees, which are almost wanting in these
forests.   Of the flowering plants that I have found
here, one of the most showy is a plant of the
Justicia tribe (not however a true Justicia), with
rose-coloured blossoms four inches long.   Passiflora
alata is plentiful on the heaps of stones which have
been formerly thrown up in washing for gold.   On
the whole, I observe that this neighbourhood has
many plants in common with Rio, and many others
with Villa Rica, but the species belonging to the
true campos are almost all wanting he re.

I have resolved to go on to Tejuco, which I hear

is only a week's journey from hence, and Colonel 1834
Skerrett has asked me to make another stay here
on my return. I enclose a letter to my aunt, which
I hope you will be so good as to forward.

Ever your affectionate nephew,

C. J. F. BUNBURY.

———

Gongo Soco
July 29th, 1834.

MY DEAR UNCLE,

I have been stationary here ever since I last
wrote to you, having given up the design of going on
to Tejuco, so that at present I am only waiting for
my letters from Rio, which I do not wish to miss.
Colonel and Mrs. Skerrett are very kind to me, and
my time passes very pleasantly, as the neighbourhood affords me a good deal of botanical and
geological amusement. I have lately observed some
limestone, the first I have seen in Brazil, but its
junction with the other rocks is concealed by soil
and vegetation; it is granular and mixed with mica,
and very likely primitive. I have heard a great
deal here of the Prussian botanist Sellow, who was
Colonel Skerrett's guest for two months, just before
his fatal expedition to the Rio Doce, and is described
as having been a most agreeable and accomplished
man, independently of his scientific knowledge.

The weather has been hot lately, at least for this
time of year, and some heavy showers of rain have
fallen. The rainy season is expected to set in

1834 towards the middle of September, so I must not lose much more time in setting off, as I have no mind to be hurried. I propose to visit the convent of Caraca on my way back to Villa Rica, and to return to Rio by S. Joao del Rey.

Ever your affectionate nephew,
C. J. F. BUNBURY.

———

Gongo Soco,
July 29th, 1834.

MY DEAR FÁTHER,

I daresay you will be surprised to find that I am still here, a month after the date of my last letter to you, and indeed I am myself surprised, on looking back to find how long my stay has been, for the time has passed very quickly. I had no intention at first of remaining half so long, but Colonel and Mrs. Skerrett have shewn me so much kindness, and made me so much at home and pressed me so much to stay with them as long as I could, that finding the neighbourhood very favourable for my pursuits, I have staid all this time. Indeed I am now waiting for letters from Rio, among which I hope to receive some of yours, for I am very anxious to hear how you are all going on. I do not well remember what my last letter contained, but I suppose I must have given you some account of the situation and appearance of Gongo Soco. I have collected a great number of plants new to me, in particular ferns, which are remarkably numerous 1834 here ; above all the tree ferns are the glory of these woods, being larger and more abundant than I have seen them anywhere else, and well supplying the place of the Palm trees, of which there are very few. You cannot imagine what an ornament these giant ferns are to the cool shady dells and moist recesses of the forests, indeed I think them the most beautiful of all the plants of Brazil. There are also some very fine species of passion-flower here, especially the Passiflora alata, which you may have seen in English hot-houses. One of the plants I have collected (a very insignificant one to look at) was pointed out to me as being very poisonous, and sometimes used by the negro slaves to revenge themselves upon their masters. It is a curious thing that in all the time I have been here, rambling out in every direction, and exploring the woods and hills daily for plants, I have not once seen a snake. The nuisance of this country is the quantity of carrapatos, most villainous little insects like ticks, which fasten upon one whenever one goes among the bushes, and bite abominably. There is a considerable variety of rocks within a short distance of this place—iron mica slate, talc slate, quartz rock, gneiss, and granular limestone, and the quantity of iron ore in various forms is quite surprising. I should think this would prove a more permanent advantage to the country than the gold mines. There is some fine wild scenery hereabouts, and very extensive views from various points on the mountains ; but

1834 certainly in the 400 miles I have hitherto travelled in Brazil, I have seen nothing comparable in beauty to the immediate neighbourhood of Rio de Janeiro. The scenery of the interior has in general too much sameness, and above all is deficient in water, for the numerous streams are too small to produce much effect in the landscape. I had at one time some thought of proceeding further north, to the diamond district, which is about 180 miles from this, but I have given it up, as I hear the country is very barren, and not interesting in any other way than as producing diamonds, and besides, provisions are extremely dear there. The excessive drought last year produced an absolute famine in that, and some other parts of Minas, so that, I am told great numbers of persons actually died of starvation. The rainy season is expected to begin in about six weeks from this time. I hope you are enjoying your summer in Switzerland, and that Edward has quite recovered his health. I long to hear from you.

Believe me ever,

Your most affectionate son,

C. J. F. BUNBURY.

Gongo Soco,
August 30th, 1834.

MY DEAR UNCLE,

I am very much obliged to you for your two letters of the 2nd and 15th of this month, and indeed I write chiefly to acknowledge the receipt of 1834 them, having very little to tell you. Before your last letter arrived, I had seen in the papers the account of the change which has taken place in the Ministry, and I am very glad to find that your opinion of it agrees entirely with that which I had formed. The debate of the 2nd of June appeared to me uncommonly interesting, and though I do not agree with Mr Stanley in opinion, I admire his speech on the question. As for the Buenos Ayreans, I suppose they will end, like the Kilkenny Cats, in eating one another up.

I intend to leave this in a few days for the Serra de Caraca where I expect to find a great many plants, and may probably arrive at Rio by the beginning of October. I have collected a good many plants since I last wrote, but nothing very remarkable ; there is a great variety of Compositæ, particularly climbing ones, here as well as at Rio, but, as you observe, they are very unsatisfactory plants to dry. Stifftia chrysantha I have seen nowhere but at Rio. Altogether, I have added between seventy and eighty species to my collection during the two months I have been here, of which twenty-eight are ferns. I will write again either from Villa Rica or S. Joao del Rey.

Your affectionate nephew,

C. J. F. BUNBURY.

1834

Gongo Soco,
August 30th, 1834.

MY DEAR FATHER,

I have within these few days received your two letters of March 18th and April 17th, and am very sorry to find that you had been uneasy at the non-arrival of letters from me. I wrote to you by a man of war (the Tyne if I remember right) towards the end of November, informing you of my intention of proceeding immediately to Buenos Ayres; that letter you ought to have received by the middle of February. My next letter was written from Buenos Ayres about the middle of January, and enclosed letters to Emily and Edward. The last newspapers have brought the account of a partial change in the Ministry, and a change, as it appears to me, much for the worse. I am sorry to see the Ministers so wavering and lukewarm, and courting the support of the Tories on so many points instead of pursuing a course boldly and decidedly liberal; and I am afraid that if I were in Parliament, I should seldom be able to vote with them. Above all, I dislike Lord Althorp's way of *shirking* the question of the appropriation of church property. Mr. Stanley's speech on that question, in the debate of June 2nd, is very fine though I do not agree with him in opinion. I am almost sorry that the contest in Portugal has terminated without Don Miguel being brought to justice for his crimes; banishment to England with a handsome allowance, is anything but a punishment. My friends at Buenos Ayres are again in a state of civil war, and I suppose they 1834 will end like the Kilkenny cats,—a consummation devoutedly to be wished. I intend to leave this delightfnl place in a few days in order to arrive at Rio before the great heats set in. There have been several rainy days in this month—an unusual circumstance, but with that exception the weather has been as pleasant as can possibly be imagined, perfectly clear and bright without any unpleasant degree of heat, and all nature is in full beauty. I am writing this in a room perfumed by the orange trees in full blossom before the window, with a most brilliant humming bird buzzing about the flowers. By the way, the Portuguese name for the humming bird is particularly expressive and pretty; they call it Beigaflor, the flower-kisser. I am very glad to hear that Edward is become a geologist, and I daresay a visit to Naples will confirm his taste that way. I have collected a good quantity of rock specimens and minerals in this country, but have not met with a trace of fossil remains. I am very much obliged to Edward for his letter and will write to him as soon as I return to Rio. I hope that the next letter I receive from you will give a more favourable account of your health which I was very sorry to hear was still weak; and I hope too that the next time I write I shall have more to say.

Believe me ever,
Your very affectionate son,
C. J. F. Bunbury.

1834

Gongo Soco,
October 18th, 1834.

My Dear Father,

I daresay you will be surprised to find that I am
still here, and in truth I am not quite certain that I
have done well or wisely in remaining so long in one
place; I have several times intended to leave it,
but have always been induced by Colonel Skerrtet to
stay a little longer, and thus I have lingered on and
on till now four months have passed away. In the
way of botany, my stay here has answered very well,
for in almost every walk I have picked up something
new; and as for climate, I believe it would not be
easy to find a healthier or more agreeable one;
Colonel Skerrett says that in the four years he has
been here, he has never seen the thermometer
above eighty four degrees, and seldom so high. In
the earlier part of this month, indeed, there were
several oppressive days, but the rain which has
fallen since has cooled the air, and rendered it very
pleasant, and its effect upon the vegetation has been
quite surprising. The flowers and young leaves
have burst forth as if by enchantment, so that the
other day I found five new plants in a space which
I had explored dozens of times before. In particular
the various kinds of myrtles, which abound in this
neighbourhood, and are just come into blossom, are
very ornamental. To say the truth I am growing
fond of this country, so that I really think, if it were

not for the wish to see you and some of my other 1834
English friends, I should have no objection to spend
four or five years in Brazil, in which time I should
be able to make myself pretty well acquainted with
its productions. I am not however by any means
charmed with the manners or customs of the people ;
they are very unrefined without being a bit the
more honest or sincere on that account. The ladies
are peculiarly unattractive both in their looks and
manners—bold, coarse, and masculine, with voices
admirably well suited for scolding their slaves, which
is their chief employment. I see by the last newspapers that Lord Melbourne has succeeded to Lord
Grey, and that the Coercion Bill has been brought
in without the political clauses, which last I am
glad of. Do you approve of the new Poor Law ? It
is a question requiring so much knowledge and
consideration, that I cannot venture to form a
decided opinion upon it, but it would seem, if the
newspapers may be believed, to be very unpopular,
and I see that many petitions have been presented
against it. I was interested by the account of the
Duke of Wellington's installation at Oxford ;
the Tories and High Churchmen seem to have
taken uncommon pains to shew their bigotry and
hostility to the general feeling of the nation.

I suppose by the time this reaches you, you will
be settled in the south of Italy ; but I am merely
guessing at random, for it is two months since I
have received any letters, in consequence of the
uncertainty of my plans.

1834   Pray give my best love to Emily and Edward, also to George and his girls, if they are with you.

Believe me ever,

Your most affectionate son,

C. J. F. BUNBURY.

***

He returned to England I think in the spring of 1835, and he found his father and Lady Bunbury returned from Italy. He spent his summer with them at Barton, and I believe he visited Stanney, his father's estate in Cheshire.

1835

JOURNAL.

Barton.

October 14th. I saw the Comet this evening, the first I ever saw. To the naked eye, it appeared like a large star with a long under-fixed train of light stretching upwards from it ; through an ordinary glass it looked more like the flame of a candle or lamp reflected in water.

October 18th. I have determined with my father's approbation not to accompany him to Nice, but to spend the winter in London, pursuing my historical and political studies, and mixing in society as opportunities may occur. It would perhaps be more amusing to go abroad, and I should certainly have the comfort of not being separated from my nearest and best friends, as well as of escaping the cold fogs of an English winter, with all their attendant miseries—no small consideration to me. But the reasons on the other side are much more weighty ; I should be unable to carry such a stock of books as would be requisite for following up my plan of study ; the journey and the temptations to wandering, which a new and fine country and a southern climate would throw in my way, and especially the opportunities of botanizing would almost irresistably divert my attention from more important objects ; and I

1834 should be in great danger of falling into habits of dawdling and desultory reading, to which I know I am naturally inclined. On the other hand, these evils would not be balanced by any opportunities of mixing with good society, and the time would slip away with as little improvement and permanent advantage to me as that which I spent at Rio.

November 3rd. Arrived in London.

London.

November 11th. I am now at last fairly settled in my lodgings, and hope to apply myself in good earnest to study, especially after to-morrow, when my father will have taken his departure for the continent. I was shocked yesterday to read of the death of Lord Milton—so young, with life just opening upon him under such favourable auspices, with every thing to make him happy!

Of my few Cambridge acquaintances—for I had fewer than almost any man of my station—four are dead since I left the university. This is enough to give one sad and serious thoughts.

November 12th. The fineness of the day tempted me to go to the Zoological Gardens in the Regent's Park, and though the weather soon changed for the worse, and became very cold, I staid there some time, being much amused with observing the animals.

There are many fine birds of the eagle and vulture tribe, particularly the Lammergeyer the harpy eagle of South America, the golden eagle, and the griffin vulture of the South of Europe.

The Condor, though decidedly the largest of its tribe, is a heavy stupid-looking bird, of an ignoble 1834 aspect; the griffin vulture is not much inferior in size, and a far fiercer and nobler looking bird, with the eye of an eagle.

The birds which they have here under the name of the Turkey vulture are different from the specimen so named in the British Museum, and from the Urubù of Brazil. The lion and tiger were in a considerable state of excitement, pacing their dens and roaring, and I was struck with the difference of their voices; the roar of the lion is deep, hollow, and rather short, while that of the tiger is a sort of lengthened snarl.

November 14th. Went with Sophy* to the Olympic Theatre, where four pieces were acted. The first was " The Daughter" a pretty little thing, with a very pathetic scene, which was extremely well acted. The next was " The Beau Ideal," a new farce by Mr. Lover, in which Liston was irresistably laughable as an old bachelor who fancies himself in love with an unknown charmer. The scene in which the nervous and valetudinarian old bachelor, in consequence of a message purporting to come from his fair one, goes to serenade her on a rainy evening, playing on a violincello under her window accompanied by his servant on the bugle, and is rewarded for his melody by a pailful of water thrown from the window on his head, was quite overpowering.

Thirdly came " A Gentleman in Difficulties "—

---

* Sophy Ramsbottom, afterwards Mrs. Thomson; she was the adopted daughter of C. Bunbury's great aunt Mrs. Gwyn.

1834 Liston on again, and hardly less diverting than in
the former piece. I ached all over with laughing.
The last thing was " The Beauties of Charles the
Second," a mere show but pretty enough in its way.
Got home by half-past-eleven.

November 15*th.* Spent the afternoon and evening
very pleasantly with Miss Fox, who as usual was
most kind and agreeable. She talked much about
Sir James Mackintosh, whom she represents very
much according to the idea I had formed of him—a
most benevolent and amiable man, of splendid
abilities, but irresolute, good-natured even to a
degree of weakness, and so utterly improvident and
careless about money, that if he had had a very
large fortune instead of a small one, he would have
been always in difficulties. There has been a story
current, and generally believed, that Sir James
appropriated to his own use some subscriptions
which had been entrusted to him, to be forwarded
to one of the unfortunate men who were transported
for sedition; she told me that this was not true,
which I was very glad to hear. The fact was, it
seems, that after he had received and transmitted
the whole of the first subscription, which amounted
(I think) to £150, a second subscription, was
attempted, but with very little success, only £3 or
£5 being raised. Mackintosh, with his usual carelessness about money matters, took no note of this
trifling sum, and forgot to forward it, upon which
his enemies raised a report that he had embezzled a
subscription of £500.

Dr Parr, a violent and hasty man, who was angry 1834 with Mackintosh for a supposed desertion of the Whig cause, gave credit to this story, and contributed to spread it.

Miss Fox gave me an interesting account of Mr. Faraday, the great chemist. He was the son of a journeyman blacksmith, was entirely self-educated, and has, by his own exertions, not only attained a most distinguished rank in science, to which his whole heart and soul are devoted, but also taught himself several languages, and in particular acquired the power of writing and speaking English with perfect propriety and elegance.

He belongs to a very singular sect called Sandy-manians, one of whose tenets is, that it is unlawful to accumulate money beyond what is wanted for present use; and to this tenet he adheres so con-scientiously, that he has repeatedly refused pensions and salaries, and lives contentedly with his wife on £100 a year.

Sir James South, the astonomer, called on Miss Fox in the course of the afternoon. He is an intelligent-looking man, with a frank, hearty, and rather boisterous manner, a loud voice, talks very fast, but agreeably. He glories in being a red-hot Tory, but seems a very good-humoured one, and without any bitterness or malice.

I had nearly forgotten to put down another cir-cumstance, that Miss Fox told me about Sir James Mackintosh. His second wife, Miss Allen of Cresselly, fell in love with him from reading the

1834 Vindiciæ Gallicæ before she ever saw him. She at once declared her determination never to marry any other man than the author of that pamphlet, and she carried her point.

November 18*th*. I walked down to Chelsea, and sat a good while with Dr. and Mrs. Somerville, conversing much to my satisfaction. Among other things, she told me that an extraordinary quantity of meteors had been observed in America on the night of November the 12th, 1833, the same day of the same month that Humboldt and Bonpland thirty years before saw those remarkable ones at Cumana; they were visible in very many different parts of North America, and even as far south as Jamaica. On the same day of the year 1834, some were visible, though by no means in so good a number, and again this year a remarkable flight of similar meteors was observed on the 12th of November, near London. Dr. Somerville walked with me to Somerset House, put down my name as candidate for the Geological Society, and got leave for me to attend the Meeting this evening. Accordingly, I went thither a little after eight o'clock, and staid till eleven, much interested with what I heard.

The first paper read consisted of observations on some parts of the west coast of Greenland, tending to prove that the land had sunk considerably within a comparatively short period. The next was a statement by Captain Fitzroy of the Beagle, relative to certain changes in the currents, and soundings on the west of Chili, produced by the great earthquake of February last; the third, an 1834 account of the same earthquake, in a letter from a gentleman at Valparaiso to the President of the Society (Mr. Lyell). It appears that the earthquake not only destroyed Concepcion, Talcahuano, and many other towns, but shattered rocks, and changed the level of the coast in a very remarkable manner: the little island of St. Maria was elevated nine feet, the coast of Concepcion three or four feet; in the port of Talcahuano, the sea suddenly retired to a considerable distance, and then returning in a tremendous wave, swept away great part of the town; the ground in many places opened in large fissures, through which water spouted up, and at the same time several volcanoes among the Andes, in particular one opposite the island of Chiloe, were in a state of eruption. Professor Sedgwick then read some interesting extracts from the letters of Mr. Darwin, relative to the Geology of Patagonia, the Falkland Islands, and the Chilian Andes.

After this, Mr. Greenough got up, and made an amusing speech of some length in his peculiar sceptical spirit, attacking the prevalent feeling of *elevation*; he spoke very fluently, and with a good deal of humour.

One or two other gentlemen also spoke, and then the meeting broke up.

November 23rd. I spent, or rather wasted, the evening from dinner, till half-past ten, at the Club, lounging over Pelham and the New Monthly. That club, I perceive, though a great convenience in many

1834 respects, is likely to draw me into habits of idling and desultory reading, against which I must be on my guard.

November 26th. I dined at Little Holland House, where there was rather a large party,—Lord Lansdowne, Colonel Fox, Colonel Mitchell, Mr. Faraday, Sir James South, Mr. Smith and Dr. Holland; there was much miscellaneous conversation, but I do not recollect anything very remarkable. Sir J. South talked with most merciless rapidity and perseverance, throwing all the rest of the company quite in the background. The moment he had quitted the room, everybody burst into an unanimous exclamation against him as a tremendous bore. I was much amused with Colonel Fox's vehemence on the subject. Lord Lansdowne mentioned a manuscript which Jeremy Bentham had shewn him, giving an account of his (Bentham's) own life; one chapter of it was headed thus :—"The ladies make love to Jeremy Bentham." Mr. Faraday, the great chemical philosopher, is a middle-sized, and very dark man, with an animated, expressive, clever countenance, and great vivacity of manner.

November 27th. I spent this day quietly at Little Holland House, and dined tête-à-tête with Miss Fox. She gave me an account of her meeting Coleridge, at Bowood, some years ago: soon after the beginning of dinner, some one happened to mention the literature of Germany, upon which Coleridge launched into a dissertation on that subject, and continued to hold forth during the whole time of dinner without pause or interruption, allowing no one else to get in a word. After he had joined the ladies in the drawing-room, he volunteered to repeat Christabel, which was then unpublished, and went through the whole of it without stopping. He used to prime himself for society by large doses of laudanum. Miss Fox says that Jeremy Bentham's manners were quite unlike those of any-one else, but there was a singular charm in them, they were so perfectly natural and simple. I should have mentioned that I accompanied her on a visit to Mrs. Calcott,* a very agreeable and remarkable woman, whose conversation in spite of her infirmities, was full of cleverness, vivacity and information.

LETTERS.

4, King Street,
November 29th, 1835.

My DEAR FATHER,

I received a long letter yesterday from my dear aunt, who says that her headaches are decidedly better.

I have seen the Somervilles twice since you went.

Since you went, I have read Mackintosh's Vindiciæ Gallicæ, and Paine's Rights of Man, and a good deal of Montesquieu's Esprit des Lois, which interests and pleases me very much, though I think his admiration for the Republics of antiquity quite over-

* Wife of the celebrated artist, Augustus Calcott, and herself an authoress of great ability. As Mrs. Graham she was governess to the young Queen of Portugal.

1835 strained, and through his love of epigrammatic conciseness often makes him obscure, and sometimes to sacrifice truth to point. I am delighted with the leads him Vindiciæ, and particularly with the demonstration that church property is public property. The plan I have followed for some time past of keeping a journal of my studies, with abstracts of what I read, and observations, takes up a good deal of my time, but I do not think it is time thrown away.

Dr. Somerville has proposed me for a member of the Geological Society, and got admission for me to one of their Meetings, where I heard some interesting papers on South American geology, and on the effects of the earthquake of February last in Chili. Mr. Greenough made a speech of some length, in his usual spirit of scepticism, attacking the Lyellian theory, with a good deal of humour, but not much argument; he seems to have become as much wedded to his doubts, as other men to their theories.

I have seen Mrs. Gwyn again to-day, and find her rather better, and more placid, though still in low spirits. Sophy and I went one night to the Olympic Theatre, and had a good laugh at Liston.

I presume you will learn all public news from Galignani, so I need not tell you of the fire at Hatfield.

Pray give my love to Emily, and thank her for the letter of introduction she sent me.

Your affectionate son,

C. J. F. BUNBURY.

London.
November 30th. 1835

I dined at Holland House; the party consisted of Miss Fox, Sydney Smith, Mr. Rogers, Mr. Cutlar Ferguson, Mr. Thompson and Sir W. Parker, besides Lord and Lady Holland and Mr. Allen.

There was much agreeable conversation: Sydney Smith very entertaining; he maintained that *absent* men were generally very happy, which the rest of the company would not admit. An animated discussion upon India, chiefly between Mr Fergusson and Mr. Smith, Lord Holland and Mr. Allen occasionally taking part.

Sydney Smith contended that the English government in India was not preferable to that of native princes, and quoted the opinions of Lord W. Bentinck and Captain Burnes on his side; Mr. Ferguson maintained the contrary. Lord Holland's frank, good-humoured manner, readiness to' talk, and fund of agreeable anecdote, gave a very pleasant tone to the conversation. He told many anecdotes of Gilbert Wakefield, whom he knew well.

December 2nd. Rather an idle day with me. I walked down to Holland House, saw Miss Fox, and went with her to call on Mrs. Calcott, who shewed me her beautiful drawings of Brazilian plants, many of them new to me. Her conversation is very agreeable; she has a surprising variety of information, which she communicates in a very unaffected and pleasing manner, and in admirable language.

1834 She has a fine intellectual countenance, the expression of which is probably softened by time and sickness, for it is much less harsh and masculine than in a portrait I have seen of her as Mrs. Graham.

December 6th.   Dined with Dr. Somerville at Chelsea, and spent a very pleasant evening. The only guests, besides myself, were Mr. Murchison, Mr. Greig, (Mrs. Somerville's son), and General———, a heavy red-faced old gentleman, who drank much and spoke little.   Mr. Murchison, one of the most distinguished members of the Geological Society, is a very gentleman-like agreeable man, with much conversation on a variety of subjects.

Mrs. Somerville delightful, uniting the greatest depth and variety of knowledge with the most unassuming simplicity and gentleness of manners, equally ready to listen or to talk, and leading the conversation without the slightest display.   I like and admire her more and more every time I see her.   The conversation never flagged, and touched upon almost all subjects, including politics, though Mr. Murchison is a Tory, and Dr. Somerville a most zealous Reformer.   Mrs. Somerville spoke of Mr. Macaulay as one of the most remarkable men she had ever met, for the brilliancy of his conversation and the wonderful power of his memory, and mentioned the high opinion that Sir James Mackintosh expressed of him.   Mr. Greig mentioned a gentleman of his acquaintance who had originally a very bad memory, and being determined to improve it, set

himself to learn Adams' Roman Antiquities by 1835
heart, and accomplished it; he has now an uncommonly good memory. We talked of Mr.
Warburton : Dr. Somerville mentioned Dr. Wollas-
ton's opinion of him—that he had the clearest,
soundest head that any man ever had. His house
is described as being in a curious state ; the window-
frames and doors not painted for fifteen years; all
the rooms cumbered with books, papers, specimens
and apparatus; not a chair to sit down on. His
only pursuits, Mr. Greig says, are now politics and
fishing. Mr. Murchison gave us an account of a
limestone quarry which is now worked near Dudley
in a very singular situation ; *under* a coal-mine.
They sank through the coal-beds, and came suddenly upon the Dudley limestone, separated from the
coal by a great *fault*, which keeps it perfectly dry.
Thus the coal is worked in the usual manner above,
and the limestone beneath it, at a very considerable
depth. He also described an ancient mine, which
he has discovered in Wales, (I do not recollect
exactly where), and supposes to have been worked
by the Romans as a *gold* mine. It consists of levels
or galleries on a grand scale (I think he said eight
feet high), carried through great veins of quartz,
containing pyrites which he conjectures to be
auriferous, no trace of any other metal is discover-
able. Dr. Somerville told us a story of himself and
Mrs. Somerville being in company with Sir James
Mackintosh in the royal park at Brussels, and so
absorbed and fascinated by Sir James' conversation,

1835 that the evening closed before they were aware, and
they were locked into the park.

December 11th. I returned this evening from St.
Anne's Hill, where I have been spending four days
pleasantly enough. Mrs. Fox* shewed me great
kindness, and I was happy to renew my acquaintance
with an old friend whom I had not seen for nineteen years. She is the finest and most amiable
old woman I know. Her age, I am told, is eighty-
nine; yet her sight and hearing are both tolerably
good, the vigour and clearness of her mind seem
unimpaired, and her conversation is lively, instructive and agreeable. She has none of the querulousness
of old age, but is always cheerful, contented, and
attentive to the wishes and feelings of others. I
was secretly disappointed at not feeling more
emotion on visiting the favourite abode of Mr. Fox.
On my return to London, I found a letter from
Henry, from Van Dieman's Land, of the date of
June the 14th.

December 12th. I drank tea with Louisa† and
Richard Napier, and sate till near midnight
with the latter, discussing political subjects.

December 13th. I walked to new London Bridge,
and across it, for the first time in my lfie; it was
too foggy for a good view of the river.

December 15th. In the evening, a large musical
party at Mrs. Gwyn's : two Miss Ravenscrofts, very
pretty girls, the elder especially; Mrs. C. Gore,

---

* Widow of Charles James Fox.

† Louisa, the half-sister, and Richard, the brother of Lady Bunbury.

Mrs. Collett with her little girl, a very pretty 1835
engaging child who played on the pianoforte and
sang with unembarrassed confidence; Mr. Lover,
who sang some capital Irish songs etc.

December 21st. Walked to the British Museum,
and spent some time in the gallery of minerals. I
was struck with the beauty of the native silver from
Kongsberg in Norway, in a variety of cord-like and
capillary shapes, fantastically twisted and curled,
and also in aggregated cubic and octohedral crystals,
more or less imperfectly formed, always accompanied by calcareous spar, and sometimes by epidote
and asbestus. The specimen of the same metal
from Peru, in dendritic and knitted forms, resembling the most delicate fern or the frostwork on a
window, are very beautiful. There is native gold
from Minas Geraes, in a great variety of forms,
dendritic disseminated in iron-mica-slate, in
curiously figured leaves, upon quartz-rock, crystal-
lized in cubes and octohedrons, and in rolled grains
of many shades of colour. There is also gold from
Sumatra and Malacca, thickly disseminated in
rolled masses of white quartz;—from Siberia, in
brown iron-ore;—from Chili, in malachite;—and
from Mexico in the form of broad flattened grains,
which look as if they had been beaten out with a
hammer.

December 23rd. Went to Drury Lane and saw
" The Jewess," a new play translated, or at least
imitated, from the French, which has had a
surprising run. Although this was the thirty-third

1836 time of acting it, the house was crammed, and I
could get only a place in the end box of the upper
tier.   It is said the receipts of this one night
amounted to £459.

It is certainly a splendid pageant, and well got
up, and I presume it is in the gorgeousness of the
show that the great attraction consists; for I do
not perceive any extraordinary merit in the play
itself—Miss Ellen Tree's acting, in the character of
Rachael the Jewess, is natural and touching.

January 11*th*.   This long interruption in my
Journal has been occasioned by a visit to Bowood,
Freshford, and Reading,* during which I have had
little opportunity of serious reading.   I was absent
from London just a fortnight.   On my return to
London, I found a long letter from dear Emily; the
affectionate and sympathizing kindness of its tone
soothed and touched me exceedingly.   She is
certainly one of the best women in the world.

————

Oxford and Cambridge Club,
January 13th, 1836.

My Dear Father,

I suppose Galignani will have given you the
general information of the glorious result of the
municipal elections, almost all through England;
it has quite surprised me, and the success of the

* Bowood—Lord Lansdowne's. Freshford—his uncle Colonel William
Napier's. Reading—Mr. Richard Napier's, these two last, brothers of Lady
Bunbury.

Liberals has been as complete at Bury as anywhere, 1836 though you know what exertions the Tories were making even before we left Barton. Of the Town Council as at first elected, sixteen were decided Reformers.

Mr. Eagle, the new Mayor, is in town, and I had some talk with him the other day; he is in high spirits as you may suppose, and says the Bury Tories are quite stunned by the change, which they did not at all expect, up to the very day of the election.

The result of the municipal elections altogether has amply shown, both how much the bill was called for, and how little mischief the Lords did after all their demonstrations. It shews, too, the folly of the violent Radicals, who would have rejected the bill, because the Lords had tampered with it, and thereby if they had done no other mischief, would have delayed this salutary change for at least another year. The Tories, as they cannot deny how thoroughly they have been beaten, want to make out that this will have no effect upon the Parliamentary elections. We shall see. Both parties, it seems, intend to muster as strong as possible on the meeting of Parliament.

I thank you for your letter of the 20th of December, which I found on my return hither from the West; I am very glad that you find Nice pleasant and healthy, and that Hanmer has joined you. Of my visit to Bowood, Freshford, and Reading, I gave Emily an account in my letter of

1836 last Sunday,* and shall only say now that it did me good in every way.

I had the happiness to see my dear aunt and cousins looking very well, and to be received by them with warm affection. My uncle William is singularly well.

I have lately been reading and thinking much about the ballot, and have at last cleared up all my doubts and difficulties on the subject, chiefly by the help of Mr. Mill's and Mr. Grote's pamphlets, so that I shall be able to pledge myself to it with perfect security of conscience. I was rather in hopes you would have told me something about French politics, but perhaps you did not remain long enough in that country to form any satisfactory judgment. Richard Napier was reading, when I was with him, the first Report of the Poor Law Commissioners, and was much delighted with it; I have not yet looked into it, but some parts that he read to me were very striking. Mrs. Gwyn, who is certainly a good deal better than she was three weeks ago, desires her best love to you, and is particularly anxious that you should send her a drawing of Hanmer, the best likeness you can make; it must be a whole length, standing, and a front face.

Pray give my best love to dear Emily and my brothers, and believe me,

Your affectionate son,

C. J. F. BUNBURY.

* This letter I have not found. F. J. B.

January 17*th*. I dined at Chelsea with the 1836 Somervilles. Mr. and Mrs. Murchison were there, and Mr. Greig. It was a very pleasant party. The Misses Somerville who are lately returned from Scotland, are very nice girls, lively, unaffected, intelligent and remarkably well informed. I like them for their unaffected love of fine scenery, and their passion for Switzerland. I do not know many girls whose conversation is more agreeable. Mr. Greig says it is impossible to exaggerate the misery of the peasantry in many parts of Ireland; indeed all accounts concur in this fact, however widely they may differ as to the causes and remedies of it.

January 19*th*. To-day after visiting Mrs. Somerville, I called on Mrs. Wilson—she is a very fascinating person, very pretty, with gentle, graceful manners, and a very sweet voice.

January 26*th*. I saw an exhibition of pictures of Van Dieman's Land scenery. It seems to be a fine picturesque country, with some very bold and craggy mountains. and pretty coves and bays; but the most striking peculiarity in the scenery is the uniform way in which a great part of the surface, both hill and dale, is covered with a sort of open wood; not thick matted forest, as in Brazil, but large trees standing at some distance apart, like the more wooded parts of an English park, without much under-wood. The trees represented in these views (chiefly Eucalyptuses or gum trees), have mostly immense trunks, and wide-spreading crooked branches, with rather thin and scanty foliage.

L

1836   Dined with Colonel and Mrs. Wilson, and spent a most agreeable evening.

February 2nd. This being a very wet day, recalled to my memory that on the same day last year I was exposed to much heavier rain in a very different situation, when crossing the Serra de Santa Anna, in the province of Rio de Janeiro. I have a lively recollection of the discomforts of that day; the exposure of four or five hours together to torrents of rain against which an india-rubber cloak afforded scarcely any protection—the toiling up and down these steep mountains in such weather at the slowest pace of a mule, through tenacious clay and over rocks, rendered doubly slippery by the rain, and the miserable evening in a barn-like building, with the rain dripping through a dozen holes in the roof, and the wind whistling through the crevices, so as to threaten the extinction of my candle.

February 3rd. Went in the evening to the meeting of the Geological Society. Mr. Murchison read a long paper on the Alluvial and Deluvial deposits, or *drift* as he called them, of Wales and certain of the midland counties of England.

February 4th. This was the day of the opening of Parliament, but the cold drizzling rain damped my curiosity, and I did not attempt to see the ceremony. The King's speech, which I read in the evening papers,. indicates some important measures : it recommends a settlement of tithes both in England and Ireland, a provision for the Irish poor, and a measure of municipal reform for Ireland, on the same principles as that which has been carried into effect 1836 in England. If this last measure be carried, it will do more than has ever yet been done to confirm the Union, and to satisfy the Irish, by making them feel themselves on an equality with their neighbours. But I fear the Lords will either reject or mutilate it.

Oxford and Cambridge Club
February 4th, 1836.

My Dear Father,

I am much obliged to you for your letter of the 26th, which I received this morning.

Old Mr. Powell called on me the other day, and we had much talk about Suffolk politics, as he was just come from that part.

I dined on Monday last with Mr. Powell, and met the Duke of Norfolk, his grandson Lord Edward Howard, Mr. Methuen, Mr. Creevey, and two or three other gentlemen; altogether it was a pleasant party.

I assure you I am by no means a thorough-going disciple of the Utilitarian Radicals, though I think there is a great deal to be learnt from their writings, repulsive as is the style of them. They are very dogmatical, too fond of vague general terms, and too much addicted to following out an abstract principle with logical strictness to its ultimate consequences, without regard to the innumerable circumstances which in practice modify and influence the results so

1834 essentially. As a mathematician would say, they neglect the *disturbing forces*. They are too much like the Laputan tailor who made coats by mathematical diagrams. It is no doubt very necessary to correct and *test* all abstract principles by frequent appeals to experience, and careful study of history; and this I think the Utilitarians neglect too much. As for the ballot, it is too great a question to be discussed in a letter, nor do I mean to maintain that it would at once and completely put a stop to all tyrannical interference, I believe no human invention could do so, but I think the ballot would go a great way, and the evil is so monstrous that it far outweighs anything which could, as far as I see, be apprehended from secret voting. I was in hopes to have found in Tocqueville some account of its practical operation in America, but have been disappointed. However, it would seem that if it be wanted at all in that country, it must be wanted for very different reasons from those which influence us. The rich, he says, so far from interfering oppressively in elections rather shun political contests, and shrink from a competition with those poorer and more ambitious, endeavouring to secure themselves by their political insignificance; whatever tyranny there is, is exercised by the all-powerful majority. A pretty choice to be sure.

I am very glad to hear that Hanmer has got reconciled to his profession. Mrs. Gwyn talks about him continually; she is complaining a good deal, poor old soul, but really does not seem to me worse than usual. The weather for some days past has 1836 been very bad. I am very well notwithstanding, and have begun to read Thier s History of the Revolution.

This club is a prodigious comfort to me.

Pray give my love to Emily, Edward, George, and his daughters.

> Believe me,
> Your very affectionate son,
> C. J. F. BUNBURY.

February 5th. The session has begun well. There was a debate last night on an amendment to the Address, moved by Sir Robert Peel, and relating to the municipal reform for Ireland; he objected to the House pledging itself to a measure *founded on the same principles* as the English and Scotch measures. The amendment was rejected by a majority of 41 in a House of 536 members. A similar amendment, however, being moved in the Lords by the Duke of Wellington, was agreed to by the Ministers, who are evidently afraid of a defeat in that House.

February 7th. I walked down this morning, in spite of the rain to Chelsea, and accompanied Colonel and Mrs. Wilson to divine service in the chapel of the college. It was a very impressive scene and well calculated to awaken solemn and devotional feelings; the French eagles and standards

1836 overhead, and the fine old time-worn soldiers who filled the chapel, listening with an air of deep attention to the preacher, naturally carried my thoughts to the tremendous scenes in which those trophies had been wrested from the enemy, and in which those veterans had received their glorious wounds, and elevated my reflections to the God of battles, Who had protected my country through such a struggle. The chaplain, Mr. Gleig, preached a very good sermon, addressing himself especially to the pensioners in a manner which I thought appropriate and striking. After church the Wilsons invited me to luncheon and I stayed till the middle of the day.

February 24th. In the evening I went to a great party at Lansdowne House, and was well amused. Most of the company were in full dress, as it was the Queen's birthday, and the multitude of rich uniforms, scarlet or blue, the gold and silver lace, with the feathers and jewels of the ladies, had a very brilliant effect. The crowd was not very great, though the party was more numerous than I should have expected so early in the season ; there were many people whom I knew, but such a scene is not favourable for much connected conversation. Still I was amused and pleased almost without knowing why. I stayed to the last, and yet got to bed before 2 o'clock, not an hour later than usual.

4, King Street,   1836
March 1st., 1836.

MY DEAR FATHER,

I received yesterday your letter of the 17th, and am very much obliged to you for explaining so fully and candidly the situation of your affairs, and my future prospects. What you have frequently said to me on the subject in conversation, and some passages in your former letters, had quite prepared me for the substance of your statement, though of course I could not know the details; it has therefore neither surprised nor dismayed me. I do not think my own tastes or habits are expensive, and I have never (at least not for some years past) expected to be a rich man, consequently I have no imaginary height to fall from and should see nothing terrific in the prospect of living on £2,000 a year, if it were to be for life, instead of for the period you mention. I am sure I need not make any professions to you; you will not suspect me of looking forward with any impatience to being master of the property. I am satisfied that you will pursue the best and wisest means to get clear of these embarrassments, and I shall be most ready to co-operate with you in any way you may point out. I am only sorry that my Brazilian crotchet should have caused you additional expense, without proportionate advantage to me,—but at any rate it has given me some experience.

Ministers appear to be quite safe as far as the House of Commons is concerned, and the great accession of strength of which the Tories boasted so

1836 much during the recess, seems to be a negative quantity. I hope you admire the course which the Ministry are pursuing with regard to Ireland; it seems to me that nothing can be better calculated to restore confidence and quiet in that country, and to make the union real and complete. It extorts praise even from the most suspicious radicals.

The line which Sir Robert Peel and Lord Stanley took on the second reading of the Irish Corporation Bill last night, seems to me a very odd one ; they did not attempt to defend the old Corporations, but instead of *reforming*, proposed to abolish them, to have no Corporations at all in Ireland, but to vest in Commissioners appointed by the Crown, all or nearly all the functions exercised by the aldermen and town councils in English boroughs. And this, as it would seem, evidently, (though not quite avowedly) because the majority of the Irish are Catholics, and *therefore* unfit to manage their own concerns. Surely if they could carry such a proposition as this it would supply the Irish with the best practical argument for the repeal of the Union. It would be a beautiful grievance for O'Connell. I am reading Thier's History of the Revolution, and have got as far as the beginning of the king's trial; it is fearfully interesting. I have kept a sharp look-out for resemblances between the early part of the French Revolution, and our own present circumstances, and I see many points of similarity, but there are also many and important differences and all, I think, in our favour.

I must tell you that Mrs. Gwyn is very unhappy 1836 at your not writing to her, but she is tolerably well in health. I was at a large party at Lansdowne House last Wednesday, and am going to another to-morrow, and to a party at Mrs. Leicester Stanhope's to-night (March 2nd). Pray give my love to Emily and to the Napiers, and believe me,

Your very affectionate son,

C. J. F. BUNBURY.

March 2nd. I went with Sophy* to a musical party at Mrs. Leicester Stanhope's. I had never seen Mrs. Stanhope before, and was rather disappointed in her beauty, of which I had heard much ; yet she is certainly a very pretty woman. Among the guests was the author of " Pelham "† whom I had had a curiosity to see, and also his wife; he is a tallish thin man, very like the portrait prefixed to the new edition of " Pelham." Mrs. Bulwer is a large woman, not a good figure, but her face is beautiful.

March 3rd. In the evening a party (called a small one) at Lansdowne House—very pleasant— a great many pretty women and agreeable men. Indeed the display of beauty was remarkable. There was an uncommonly pretty girl who I understood was Lord Erskine's daughter ; her figure and dress gave her quite the look of an old picture.

* The adopted daughter of Mrs. Gwyn.
† Edward Bulwer, afterwards Lord Lytton.

1835   *March* 13*th*.   I went this afternoon to St. James'
church, the third time I have been to church this
year.   But the service had not long begun before I
found myself getting into a better and holier frame
of mind.   There is certainly something in the
solemnity of the scene, in the spectacle of a great
multitude, of all ages and classes and conditions,
assembled for the purpose of prayer and praise,
listening with apparent earnestness and reverence
to the appointed service of the national Church, and
offering up with one accord their confessions and
petitions to heaven, there is something in all this
which has a strong tendency to awaken devotional
feelings.

April 5*th*.   Here is a long gap in my journal, of
which I cannot give a very satisfactory account.   I
went down to Barton, on the 19th of March, and
stayed till the 31st, employing almost the whole of
my time in the examination and arrangement of my
botanical collections.   I was rather pleased to find
that after having so long devoted my attention to
politics, I could transfer it so easily and completely
to a quite different subject.   I returned to London
on the 31st.

May 5*th*.   I was at Little Holland House from
1 till 4 o'clock, many visitors came in.   One of
them was the Rev. Mr. Stanley,[*] who talked about
the Hampden controversy, and told some anecdotes
of the bigotry of the orthodox party at Oxford.   In

---

[*] Afterwards Bishop of Norwich

particular, he mentioned that his son,* who is an 1836
undergraduate there, sent in an English essay on
some subject quite unconnected with the controversy; the Master of Baliol objected to the word
" Facts " which occurred in the essay, for, he said,
" that word had lately been used in a very unfortunate sense, and was therefore highly objectionable."
This was an illusion to the title of a pamphlet
published on Dr. Hampden's side.

Went to a rout at Lansdowne House in the
evening; the whole suite of rooms was thrown open,
and crowded to excess, and the heat was great.

May 6th. Dined at Little Holland House, with
Lady Mary Fox, Lady Dacre, Mr. Calcott,† and
Mr. Vernon Smith. Lady Dacre is a very animated,
clever, agreeable person. Mr. Calcott has a
remarkably mild gentlemanlike manner, with a
quiet dry humour that is very amusing. Sergeant
Talfourd's tragedy of " Ion " was talked of.
Mr. Smith said that he figured the author
to himself as a sort of " Chimaera,—an unnatural
combination of lawyer and poet,—Melpomene in
a sergeant's wig." He does not think very highly
of that tragedy. Sergeant Talfourd is said to be a
prodigious admirer of Coleridge, and given to
holding forth in company in Coleridge's manner. I
was surprised to find that all the three ladies present
were against the project of admitting ladies to the
gallery of the House of Commons.

* Afterwards Dean of Westminster.
† Afterwards Sir Augustu Calcott

1836   *May 9th.* I went at 2 o'clock to the Linnean
Society, and looked over Humboldt and Bonpland's
'' Plantes Equinoxiales,'' a fine work; the plates are
on a very large scale, finely engraved, and have all the
appearance of great accuracy, but are uncoloured,
and contain no very detailed dissections.   The
accounts of some of the plants are very full and
interesting,—as for instance of the Wax Palm,
(Ceroxylon andicola), the tallest of the tribe, which
grows only on the Andes of Quindiu, at a greater
elevation above the sea than any other palm, and of
which the trunk is covered with a concretion
resembling beeswax, and used for the same purposes;
—the Cinchona Condaminea, peculiar to the neighbourhood of Loxa, and which yields the finest kind
of Peruvian Bark ;—the Bertholletia excelsa, which
produces the Brazil nuts,—and others.   Two species
of Bamboo are here figured and described,—the
Bambusa Gradua, which forms entire forests among
the Andes, especially in the pass of Quindiu,—and
the Bambusa latifolia, which grows on the banks of
the Orinoco and its tributaries.   The latter, in the
form and disposition of its leaves, resemble the great
bamboo or Taquarra of Brazil.   It is curious that,
in all the time I was in Brazil, where the forests are
full of Bamboos, of two or three kinds, I never could
find any of them in flower.   Humboldt says that
siliceous concretions (Tabasheer) are sometimes met
with in the joints of the Bambusa Guadua ; I never
could meet with any in the Brazilian bamboos.   Mr.
Don told me that the tallest tree in the world is the

Taxodïum sempervirens, a native of California, which 1836
has been seen of the height (by measurement) of 300
feet. The Deciduous Cypress, which is another
species of Taxodïum, though less lofty, grows to a
greater bulk than any other tree except the Adansonia.
The tallest of the Tree Ferns is the Alsophila
Brunoniäna, a native of the mountains of Silhet,
which is 45 feet high. Tree ferns extend into much
colder climates in the southern hemisphere than in
the northern ; they are found in Van Dieman's Land,
and in the southern part of New Zealand, but
Madeira is their furthest limit to the north.

May 11th. The first really warm day I have felt
since last September. At the Linnean Society, I
looked through Mr. Lambert's great work on the
genus Pinus, and read Humboldt's very interesting
preface to the " Nova Genera " etc. I was acquainted
with most of his general conclusions from other
sources, but many of the details were new and
interesting to me. He found, on the mountains of
South America, one species of Taxus, but no other
Coniferae, whereas in Mexico the mountains are
covered with forests of Pines and similar trees,
including species of Cypress and Juniper; the
Taxodium distichum also, or Deciduous Cypress, is
a native of Mexico, as well as of North America.
Social plants are chiefly characteristic of temperate
climates ; there are very few in the low districts of
tropical America (the Mangrove, Bamboo, and two
or three others), but they become more numerous
as you ascend to the higher regions of the Andes.

1836 The number of genera, in proportion to the number of species, is greater in temperate or cold countries than in hot ones, and on high mountains than in the plains. The species common to tropical America with other countries are chiefly grasses, cyperaceae, mosses, and lichens, with a few ferns and lycopodia, but scarcely any exogenous plants, except such as have evidently been introduced. On the higher regions of the Andes, towards the limits of perpetual snow, grasses abound beyond all other plants; whereas, on the high Alps of Europe, there are comparatively few of them. Humboldt gives a very interesting account of the three great zones, called the hot, temperate, and cold regions, which are observable in equinoctial America, according to the elevation above the level of the sea, and gives lists of the most characteristic plants of each.

In the evening I attended a meeting of the Geological Society. Mr. Murchison read a long and elaborate account of the coal-field of Dudley in Staffordshire.

May 15th. There was this day a remarkable annular eclipse of the sun which was at its height soon after 3 p.m. The early part of the day was rather cloudy, but towards 2 o'clock it cleared, and became very favourable for observing the eclipse. The darkness, however, when at its height was not near so great as I had expected, but pretty much like that of a clear day in mid winter.

In the summer of 1836 Charles Bunbury and his brother Edward made a tour through Ireland, of

which the account will be given in the following 1836
letters.

LETTERS.

Gresham Hotel,
Saturday, August 13th, 1836.

MY DEAR EMILY,

I have now been here four days, and certainly, so
far, I have every reason to speak well of Dublin, for
I have enjoyed myself very much; and never were
people more fortunate in point of weather than we
have been; this is the fifth day of almost unclouded
sunshine. We left Liverpool on Monday evening,
and entered Dublin Bay, after a very favourable
passage, on the finest of all fine mornings. I was
very much pleased with the view of the coast as we
approached,—I had heard much of its beauty, and
it fully came up to my expectations. We spent all
that day in walking about the town in all directions
and driving in a car about the Phœnix park, which
is the most beautiful and enjoyable thing of the
kind I ever saw; and the next day we called upon
Sir Guy and Lady Campbell. They have shewn
us the greatest kindness, and done everything to
make our stay in Dublin pleasant to us, and I need
not tell *you* how capable Lady Campbell is of
succeeding in this. She is certainly the most
agreeable person in the world. Sir Guy has been

1836 very good-natured and friendly. Their children are
lovely. Henry FitzRoy and his wife are here too,
and have been very kind to us. Mrs. FitzRoy is
a very pretty woman, and remarkably frank and
lively and pleasant, reminding me much of Caroline
Henry,* and she claimed cousinship with us at once
so frankly, that we were very soon on the footing of
old acquaintances. She shewed us yesterday the
Chief Secretary's house and gardens in the Phœnix
park. I have seldom seen a prettier spot. Indeed
the neighbourhood of Dublin, as far as I have seen,
appears to be very enioyable, and has immense
advantages over that of London, the views along
the coast to Howth on the one side and the
Wicklow mountains on the other, are beautiful,
and it is a delight to get out so quickly into
the country from a great city, without those eternal
and intolerable cockneyfied suburbs. The prevalence of white houses, and long lines of white
wall along the coast, as well as the look and
dress and manner of the lower class of people
remind one much of Italy. I am rather disappointed with the town of Dublin itself, in point
of architectural effect I mean ; there are some very
handsome public buildings, particularly the Post
Office, the Custom House, and the Bank, and there
are many good broad streets, but the plain brick
houses have to my eye a poor effect, compared with
Bath, or with some of the newer parts of London.

---

* Mrs. Henry Napier.

Even in the best streets, the number of ragged and 1836 barefooted people strikes me as much greater than in any large English town; I have not yet been in the worst parts of Dublin, but intend to see something of them before I leave it. We shall set off on Wednesday for the county of Wicklow; I should have liked very well to stay longer in Dublin or its neighbourhood, but the season is getting late, and we must take advantage of this beautiful weather while it lasts.

I have left myself but little room to talk of Sarah and Cecilia, and of the happy week I spent with them at Reading before we left England, but you will easily understand how delighted I was to see them again after so long an absence,* and to find them in such health and spirits and beauty, and to meet with such a kind and sisterly reception from them. They are indeed delightful girls; George may well be proud of such daughters, and you of such nieces; but I fancy you will hardly agree with me in thinking Sarah decidedly the prettiest of the two.†

It is unlucky that the Duke of Leinster is out of Ireland at present, so that I cannot deliver your letter to him, but Mrs. Fitz Roy says we ought to see Carton nevertheless.

Ever your most affectionate,

C. J. F. BUNBURY.

---

* They had just returned from Italy.
† Sarah was afterwards Mrs. Clark, and Cecilia, Mrs. Henry Bunbury.

M

1836   *August* 14*th.*   We walked yesterday through a considerable part of the " Liberties," a portion of the
city lying to the west and south-west of the Castle,
and towards Kilmainham, and exhibiting a shocking
contrast to those fine streets and squares which we
had seen before.   The filth and closeness of those
narrow streets and alleys, the dark ruinous-looking
houses, the windows with scarcely an unbroken pane
of glass, the tattered clothes hung out from the
upper windows to dry, the ragged, meagre, unshod,
unwashed inhabitants, have altogether a look of
squalid poverty that is deplorable.   At the same
time, I am so unacquainted with the poorest parts of
London that I cannot draw a comparison.   In the
Liberties and the outskirts of the city, one is often
reminded of Italy by the groups of ragged barefooted boys lying sunning themselves under the
walls.   Even in the best streets, one may often see
poor people in a like tattered condition, sitting or
lying on the steps before the finest houses.   Yet
with all this raggedness and poverty, one is very
little pestered by beggars.   I suppose this is owing
to the exertions of the Mendicity Society.

Few things catch the attention of a stranger in
Dublin more than the outside jaunting-cars, which
one sees at every turn, and which indeed take the
place of gigs and cabs, both public and private.
Very convenient vehicles they are too, though there
is something very odd in their appearance till one
becomes used to them.

I had nearly forgotten to speak of the Castle, the

seat of the executive government. It is a consider- 1836
able pile of building, but not at all remarkable for
beauty. The Irish Parliament used to sit in the
building which is now the Bank. It forms one side
of the irregular open space called College Green,
which, notwithstanding its name, has no verdure to
boast of.

August 18th. We went, three days ago, to see
Castletown, which I had so often heard of that I felt
a great curiosity about it—and certainly I was not
disappointed. It is a charming place. The park,
which abounds with fine timber, particularly beech,
is bounded by the Liffey—not here the dull, paltry,
muddy river that it is at Dublin, but a beautiful,
clear, rapid, sparkling stream, with all the character
of a mountain river, hurrying and brawling over large
stones and ledges of rock except in one part, where
being kept up by a weir, it forms a deep still pool,
reflecting with the greatest distinctness the trees and
thickets on its banks. The water is of that peculiar
rich transparent brown, like cairn-gorm, which one
sees in the Highland rivers, and its banks are fringed
with a beautiful variety of trees, of most luxuriant
growth, feathering down quite into the water. The
fine range of the Wicklow hills, which bounds the
horizon to the south, adds much to the beauty of the
scene. The house is very large, and of a massy and
respectable appearance, without any remarkable
beauty outside ; but the hall and library are very
handsome. We returned from Castletown by a very
pleasant road along the side of the Liffey, which pre-

1836 serves its wild and rapid character to within a few miles of Dublin. The banks, for some miles downwards from Lucan, are high, bold, and beautifully wooded, with many handsome gentlemen's seats, forming altogether the prettiest valley I have seen for a good while.

August 20th. We returned last night from a two days' jaunt into Wicklow, going by Bray, the Dargle, and Powerscourt, to Roundwood, and back by Enniskerry. The Dargle is strikingly like the valley of the East Lyn, in the north of Devonshire, but the stream is smaller, and the hills which rise immediately from it are less high and more uniformly wooded.

August 22nd. We left Dublin about three in the afternoon, and drove to Bray, and thence on to Newtown Mount Kennedy, where we arrived before seven.

August 23rd. It rained all the morning, so we did not leave Newtown till past one o'clock, when we set out in a car for Ashford and Rathdrum.

At Ashford, where there is a very good inn, we left the car for a while, and walked up the Devil's Glen, an excursion which well repaid us for our trouble.

This glen is a good deal in the same style as the Dargle, but very decidedly superior to it in beauty, and especially in variety, for the hills on one side, instead of being uniformly wooded, are rocky and broken, shaggy with heath and brushwood, and relieved by contrast the thick woods which clothe the opposite side. The hills too are, I think, higher than

at the Dargle, and there is rather a larger stream, 1836
which winds among the picturesque mossy rocks in
a beautiful alternation of deep still pools and foaming rapids.   We ascended to the head of the glen,
about three miles from Ashford, where the river
(the Vartrey) quitting the high moorlands, throws
itself in a very pretty cascade down the rocks.   This
is a beautiful wild spot, with fine massy grey rocks,
and purple heath, and feathery birch and ash trees,
and the foaming stream flinging itself down from rock
to rock in a succession of small but picturesque falls.
We dined at Ashford, and went on in the evening to
Rathdrum, catching a glimpse of the town of Wicklow
on our way.

August 24th.   From Rathdrum we drove to Glendalough, the famous valley of the Seven Churches,
(the glen of the two lakes).   We presently came in
sight of the principal ruins, and the singular round
tower rising above them, with the two little lakes
beyond, and the dark mountains frowning over them.

The scene was very wild and striking.   While
Edward stopped to make a sketch from this point,
I proceeded to the ruins and the lakes with a most
communicative guide, a strange-looking fellow with a
great red beard, who poured forth with prodigious
volubility a great store of legends concerning the
churches, the round tower, St. Kevin, King O'Toole,
the Danes, and the giant Fin Mac Coul " who went
to school with the Prophet Jeremiah," and who
moreover cut one of these mountains in two with a
blow of his sword.   The three principal of the Seven

836 Churches are near the round tower, just below the lower lake ; two others are lower down the valley, and the other two near the upper lake ; they are all on a small scale, and of a rude appearance, though the arches shew considerable skill, and if their antiquity be really what is supposed, they must have been well and solidly built to last so long. They are built of mica slate, the material of the surrounding mountains, but the stones forming the arches are granite. The largest of the churches, called the Cathedral, is . roofless, but its walls are in tolerable repair ; here the guide shewed me a stone in the wall with a rude carving of a wolf devouring a serpent, and related how, while the church was building, a serpent used to come out of the lake every night, and destroy all that had been built during the day, and how St. Kevin made a tame wolf or a wolf dog (I am not sure which), that belonged to him, devour the Serpent. Another church hard by, known by the name of St. Kevin's kitchen, is much the most perfect of the seven, having its roof, which is singularly high, in good repair, and a curious little round tower with a conical top in the place of a steeple at one end of it. Here mass used to be celebrated till lately, but I was told that the Priest had now forbidden it, because great numbers of people used to assemble on the occasion, and as soon as mass was over the " boys " would fall to drinking and fighting. The third church in this group, which was dedicated to the Virgin, is a very picturesque ivy-covered ruin. High above all these rises the mysterious round

tower, said to be 110 feet high and 52 round, rudely 1836
but solidly built, of unknown antiquity and doubtful
purpose, and interesting from this very obscurity.
My guide indeed seemed to be troubled with no
doubts concerning this tower, which he asserted was
built by the Pagans, " worshippers of Baal," for the
purpose of watching from its summit the rising of
the sun. I have not hitherto felt any interest in
Irish antiquities, but there certainly is something
exciting to the imagination in the singular group
of relics of such a remote age, placed in the middle
of a scene so wild and lonely, and seeming to
attest the former existence of a numerous, religious,
and to a certain degree civilized people, in a
district apparently little fitted to support them.
I have not room for many of the stories which
my guide told me, such as the extraordinary
powers attributed to an ancient stone cross,
near the cathedral, the *fecundating* virtue of St.
Kevin's bed, and the deer stone, a peculiar object of
veneration, round which the peasantry crawl on their
knees upon the festival of St. Kevin, the 3rd of June.
Around the cathedral and round tower is a modern
burying ground, crowded with graves, and hither
people are brought from considerable distances for
burial, as the place is of peculiar sanctity, St. Kevin
having prayed (so the legend says) that all who were
buried here might be saved at the Day of Judgment.

But it is time to say something of Glendalough
itself, which would be well worth visiting, even without the churches. The lower lake is little more than

1836 a pool, with swampy and reedy margins, but the upper one is a beautiful little sheet of water, about a mile long, singularly dark, deep and clear, surrounded by rocky mountains, which rise very steeply, and in many places precipitously, from its margin, so that a boat can go quite up to the rocks. It reminded me a good deal of the Capelberrig lakes, but the mountains around, wild and abrupt as they are, are far less bold and varied in their outlines than those of the Snowdon group. Their rounded tops are covered with heath, but on their sides the mica-slate shews itself in bold cliffs and sharp projecting ledges and crags. In the face of one of the most abrupt of these cliffs, about thirty feet above the water, is the small excavation (either wholly or partly artificial) called St. Kevin's bed. My guide did not fail to repeat the well-known legend of St. Kevin and "the beauteous lady called Cathleen, supposed to be a daughter of King O'Toole," whom the saint treated so scurvily; he moreover repeated "Misther Tommy Moore's" melody on the subject, with an accen and rhythm which particularly improved the effect of it. Mr. Moore would hardly have recognised his own verses. Edward and I took a boat, landed nearly under St. Kevin's Bed, and scrambled up to it with some difficulty, with the help of the woman who acts as guide to it. The water of the lake is extremely clear, but of so dark a colour that it appears quite black in looking down on it. We were very lucky in our day, for there was a bright sunshine, with flying clouds which threw strong shadows

over the mountains, and shewed their prominent 1836
points and diversities of form and colour with fine
effect. Having seen Glendalough to our satisfaction,
we proceeded by the military road over the mountains
to the inn at the entrance of Glenmalure, where we
dined. We took a glance at Glenmalure, a narrow
valley between bare, rugged, dusky mountains, much
resembling some of the Highland glens. From
thence we went on down the pretty wooded vale of
the Avonbeg, similar in character to that which we
had ascended from Rathdrum to Glendalough ; and
soon after passing a place called Ballinaclash we
reached the celebrated Meeting of the Waters,
where the Avonmore and Avonbeg unite, and take
the name of the Avoca. The junction of two clear
swift streams, with richly wooded hills on each side,
and a castellated mansion situated on one of these
hills just overlooking the point of junction, forms a
very pleasing scene. The valley widens a little
below this, but retains its green and wooded
character, except in one or two spots, where large
copper mines worked on the hills have grievously
marred its beauty. We stopped for the night at the
Wooden Bridge Inn, situated near the junction of
the Derry with the Avoca.

August 25th. Spent this day at the Wooden
Bridge Inn, and walked up one of the side valleys to
what is called the old gold mine, a spot, where, some
years ago, a considerable quantity of gold was found
in the bed of the stream, and where particles of that
metal are still occasionally met with. We bought

1836 two specimens of it from some girls who had just
before collected them in the sands of the rivulet. It
is very remarkable that the researches which have
been made with a view to discover gold in the rocks
from which this stream takes its rise, and from which
it might be supposed to derive the precious metal,
have been entirely unsuccessful. The wild wooded
dell, down which it flows from the mountain of
Croghan Kinshela, is very pretty, and extremely like
some of those in the north of Devonshire, and on the
skirts of Dartmoor.

August 26*th*. We left the Wooden Bridge Inn by
the coach soon after noon, and passed through
Arklow, Gorey, Ferns, and Enniscorthy, to Wexford,
a journey of thirty-eight Irish miles. In the forenoon
I walked some way both up and down the Vale of
Avoca. It is indeed a lovely valley, in a sweet and
gentle style of beauty. The brown, transparent,
murmuring stream is bordered by meadows of the
freshest and brightest green, and the hills which rise
in steep slopes on each side are clothed with
luxuriant woods of oak, ash and birch, which in some
places feather down to the very edge of the water.
In the evening we had a curious specimen of Irish
manners, in a party of gentlemen in the public room
of the hotel, · who became uproarious over their
whiskey, sang Orange songs in full chorus, and
toasted " the glorious, pious and immortal memory "
with great vehemence.

1836

LETTERS.

Woodstock, near Ennistioge,
August 31st, 1836.

My Dear Father,

We are staying here with Mr. and Lady Louisa Tighe, to whom Lady Campbell gave us a letter, and who have shown us great hospitality and kindness. They are very pleasant people, and are universally well spoken of for their attention and kindness to their tenantry. Lady Louisa fully confirms Emily's favourable opinion of the Irish peasantry, as a well disposed and grateful people, and she says that she meets with no opposition or hostility on the part even of the Catholic priests, but that on the contrary they approve and sanction the attendance of the Catholic children at her school. The domain is magnificent, I do not recollect to have seen such an extent of wood, and of such noble growth, near any other gentleman's house; and it has all the advantage of situation to set it off, rising up a high and steep bank from the broad fine river which winds through the valley. Then the bright green meadows

1836 on the other margin of the river, and the neat village at the bottom of the valley, with its handsome stone bridge and the high hills in the background, complete the picture. The house itself is not handsome, but the garden is beautiful, and the place altogether would make Emily wild if she saw it. An old friend of hers—Lord Devon—is here on a visit with two of his sons.

We are going to-morrow to see Kilkenny, and dine with Sir John Power at Kilfane, and the next day we shall go to Waterford, where I hope to hear of you. We left Dublin on Monday week, the 22nd, and I was quite sorry to leave it, for the kindness of Sir Guy and Lady Campbell had made our stay there extremely pleasant, and we were in the way of getting acquainted with many agreeable people. I certainly intend (*Deo volente*) to return thither before I leave Ireland. We spent three days in seeing the wonders of Wicklow. I was much pleased on the whole with that county; there is nothing grand in the scenery, nor any of that astonishing beauty which one sometimes meets with; but many wild picturesque wooded glens, and rapid mountain streams, and still, dark lakes embosomed in the wild hills, much of that mixture of grey rock and purple heath, and feathery birch tree which is always beautiful. The mountains, though nearly as high as those in North Wales, have much less variety and boldness of outline. The vale of Avoca, in which we spent a whole day, is very lovely, in a quiet and gentle style of beauty. We stayed a day at Wexford, a vile, ill-

built town, but well situated, and seemingly rather 1836
flourishing; then came on to New Ross on Sunday,
and on Monday hither. We have been tolerably
fortunate hitherto, in point of weather, and I am told
this is decidedly the best season for travelling in
Ireland. It would of course be rash to draw any,
decided political conclusions from the little I have
yet seen of this country; but I begin to see that
there are greater difficulties in the way of a satisfactory settlement than I was quite aware of. A
conversation I had at Dublin with Mr. Crampton, a
clever and sensible man, gave me rather a new view
of the subject; he remarked that the problem which
the Irish Government had to solve, was that of
reconciling and amalgamating two distinct and
hostile races of men living under the same government, the one superior in numbers, the other in
wealth, industry, and determination; and both like
two bull dogs, only waiting to be let loose at each
other's throats. This, if it be correct, is a fearful
picture. As for the state of the people, I confess I
have not hitherto, except in the Liberties of Dublin,
seen so much outward appearance of misery as I
expected, but I believe Wicklow and Wexford *are*
two of the most prosperous counties in Ireland.
Some very wretched hovels I certainly have seen, and
a considerable number of miserable looking beggars,
particularly at New Ross, and altogether there is
worse cultivation, and less appearance of neatness
and industry than, in most parts of England, but by
all accounts I shall see much worse things in

1836 Munster.   I was very sorry though not surprised to hear of poor Kerry's* death.   Pray give my best love to Emily.

Believe me ever,

Your affectionate son,

C. J. F. BUNBURY.

Cork.

September 14th, 1836.

MY DEAR EMILY,

I wrote my father a long letter from Woodstock on the last day of August, so this present epistle is for you, more especially as I know that Edward has just written to my father.   We spent five days at Woodstock, and most agreeable ones, for it is a lovely place, and nothing could exceed the kindness and hospitality of Mr. and Lady Louisa Tighe. Then we came on hither by Waterford, Clonmell, Cashel, Lismore, and Fermoy, and arrived here the day before yesterday, just three weeks after leaving Dublin.   We have now seen a good bit of the South of Ireland, and seen it in a leisurely and satisfactory way, and on the whole I have much enjoyed my tour hitherto, though we had a week of very bad weather (beginning the day we left Woodstock) which was a terrible bore.   We spent two very pleasant days

---

* The Earl of Kerry, eldest son of the Marquis of Lansdowne.

however in spite of the rain, with Billy* at Cashel, 1836 and he got leave to accompany us to Lismore and Fermoy. I like him uncommonly. The ruins on the rock of Cashel are very picturesque and interest· ing, much more extensive and massy than I had at all expected, and the view from the top of them is beautiful, but the city itself is the very worst, poorest, dirtiest and vilest town (to be called a town) that we have yet seen in Ireland, and a terribly large proportion of its inhabitants seem to be in a state of squalid poverty. A great part of the town consists of long straggling rows of the most wretched mud cabins, such as make one sick to think that human beings should be obliged to live in them. Certainly the county of Tipperary, as far as we have seen (and we have seen a large portion of it) is much the worst part of Ireland through which we have passed. From the time we entered that county, at Carrick on Suir, there was a manifest change for the worse in the habitations and dress and apparent condition of the people, which continued till we quitted it again between Clogheen and Lismore, but the worst of all was at Cashel, and in its neighbourhood. It was a most cheering and agreeable change when we crossed the Knockmealdown Mountains, and entered upon the Duke of Devonshire's estate near Lismore, where we saw neat, substantial, well-built cottages, and every symptom of prosperity. Surely the evils of Ireland are not owing to absentee-

* William Napier, youngest son of Sir George Napier. Now General William Napier.

1836 ism, for it seems to be admitted on all hands that three of the most flourishing and best managed estates in the whole country are those of the Duke of Devonshire, Lord Fitzwilliam, and Lord Hertford —three great absentee proprietors. On the other hand it must be owned that Woodstock is a beautiful example of what an Irish estate may be under the management of a really good resident landlord. Although I am afraid I see my way through the Irish question rather less clearly than I thought I did before—that is as far as legislative interference is concerned, for I see more plainly than ever how much may be done by private exertions. What strikes me most of all is the extreme difference of the Irish (I mean the lower classes) from the English—difference in features and air, and gait, and habits, and tastes, and in the whole of their character. They seem to be essentially a *southern* people, and one wonders to find them in such a parallel of latitude. And the difference of character, as it makes it more difficult for an Englishman to understand them, necessarily increases the difficulty of legislating for them. I was told by an *Irishman* whom I met at Wexford, that the feeling of gratitude was unknown to the lower classes in Ireland ; Lady L. Tighe on the other hand, speaks as highly of their grateful and kindly nature as you have always done. Is it that the Irish are called ungrateful by those who have never given them any cause for gratitude ? By the way, that Irishman we met at Wexford was a curious specimen of his kind, very

conversible, pleasant and well-informed, and on all 1836
other subjects shrewd and sensible, but his Orange
vehemence, the rancour and fury of his religious zeal,
and his abhorrence of the Catholics, were something
that I could hardly have believed possible.

I do not think it necessary to apologise for this
long grave prose about Irish affairs, for I know you
do not think politics (and least of all Irish politics)
a bore, whatever you may think of my *crudities* on
the subject. I believe we must give up all thoughts
of Connemara and the North for this time, for it is
already late, and we have several good introductions
for the neighbourhood of Limerick which I should be
very sorry not to take advantage of. Probably therefore after leaving that city, we shall proceed merely to
Ballinasloe and Athlone, and then strike across to
Dublin. But I certainly hope and intend to revisit
Ireland, and complete the tour of it some other
time. My father seems, by what Edward tells me,
to be (or to have been at the time he wrote that
letter) much alarmed about politics—I do not quite
understand why. Before the end of October I hope
to be at Barton, and to have much talk with you
about Ireland.

Yours very affectionately,
C. J. F. BUNBURY.

1836

EXTRACTS FROM JOURNAL.

September 5th. I had been told at Wexford (by an Irishman too) that the feeling of gratitude was unknown to the lower classes in Ireland ; but Lady Louisa Tighe declared there was not a kinder or more grateful people in the world if well treated. She does not meet with any hinderance or opposition even from the Roman Catholic priests,—on the contrary they sanction and approve of the school under her direction, which is attended by Catholic and Protestant children alike. She makes no distinction between the sects, and aims only at teaching the children what may be useful to them, and fitting them for good and moral and well-conducted members of society, instead of inculcating sectarian doctrines, or seeking to make converts under the guise of education, or teaching them to hate or despise their fellow creatures for a difference of opinion. Surely a school conducted on such principles must tend, in its own sphere, to cherish peace and goodwill, and soften that bitterness of religious animosity which is the plague of Ireland.

Mr. Tighe told me that the usual rate of wages in this neighbourhood was 10d. a day, which, considering the price of provisions, goes as far as 15d. in England. For task work, at particular seasons, the wages are much higher ; he pays, for mowing his grass, at the rate of 4s. 6d. per Irish acre, and a 1836 good labourer will do as much as this in the day. My brother remarked, in conversation with him, that there was not generally, in this part of Ireland, so much appearance of wretchedness and squalid poverty as he had been led to expect. Mr. Tighe said that there was not in reality a great deal of it, at least in comparison with the more westerly parts of Ireland; and he added—"we are as much superior to the West and South West, as the best parts of England are to us." The agriculture, however, appears generally to be slovenly and bad; the fields are full of weeds, and there is very little variety of crops,—scarcely anything indeed but potatoes and oats. Mr. Tighe has taken great pains to introduce an improved system of farming, and the culture of turnips, but has found the farmers very averse to innovation and experiment. This is an instance in which one may see the advantage of large farms, or of those agricultural schools which Captain Kennedy has proposed to establish.

September 7th. Went from Waterford to Carrick-on-Suir, passing through Curraghmore park, and then on to Clonmel. We passed through the village of Portlaw, which appears comfortable and flourish-ing, and is beautifully situated, backed by extensive oak woods, and by a bold range of brown heathy hills. There is an extensive manufactory here, established by a Scotchman, to which the village owes its prosperity. A little way further on is the gate of Lord Waterford's place, Curraghmore; a

1836 fine full clear stream flows through the park, and a rich bank of wood rises from it; the timber is magnificent, particularly the Spanish chestnuts, and many of the oak trees I think are equal in spread or branches, though not in girth, to any I have seen in England. The thorns, as usual in Irish parks, are beautiful. A fine range of mountains (the Commeragh), dusky and steep and rugged, forming the background towards the South, enhances very much the beauty of the park.

September 12*th.* In Cork, as in Dublin, there are strange and shocking discrepancies in the population; in the principal streets there are multitudes of well-dressed and of busy people, but one also sees, in the very same streets, a terrible number of beggars, and of ragged and miserable beings,—women and children hung with filthy tatters that scarcely serve the purposes of decency, and ragged men lounging about at the corners of the streets or on the bridges, or lying asleep on the pavement. I am every day more struck with the difference of the lower classes of Irish from the English, and their similarity to the people of Southern Europe.

September 13*th.* We spent the greater part of this day in seeing the beauties of Cork harbour. It is indeed a beautiful harbour—I do not recollect any that comes near it in its own peculiar style. It has no grand or bold features; but its numerous branches and inlets, and the islands of various sizes, and the way in which the points and headlands run out on both sides, crossing one another as it were,

and appearing one beyond another in different 1836
gradations of distance and colouring, produce an
extraordinary variety. The banks are rich and
smiling, sometimes low, more often swelling into
green hills, variegated with wood and meadow,
and thickly dotted everywhere with gay country
houses and ornamented cottages, with neat
white church-spires here and there appearing in
strong relief against the woods, while the multitude
of vessels of all sizes moving over the smooth water
give additional liveliness and beauty to the scene.
It strikes me as something like Falmouth harbour,
but it has more branches and much greater variety,
and the banks are lower and richer, and more
wooded, and altogether of a softer and tamer
character than those of Falmouth.

September 15*th*. From Cork to Macroom,—rather an
interesting journey. The quarter of Cork by which one
goes out on this road is handsome, and worthy of a
great city : there is a fine public walk, and a handsome
bridge, and the county gaol, a stately building with
a Doric portico. For several miles the road runs
along the valley of the Lee, and traverses a fine, rich,
smiling country ; then the river makes a sudden
bend, and the scenery takes a wilder character, without having much of picturesqueness or beauty; but
it improves again a few miles before reaching
Macroom. The surrounding country is hilly, and
very pretty, with a good deal of wood, and masses of
grey rock, and a winding river, and a ruined tower or
two, and several successive ranges of hills in the

1836 background. The soil seems to be poor, the rock
being generally very near the surface, and often
covered only with bog ; and the cabins are mostly
small and wretched, built not of mud but of loose
stones, very often without a window, and with
nothing but a hole by way of chimney. At a little
distance it is not easy to distinguish them from the
moss-grown rocks among which they are scattered.
It was on one of these rocks that I for the first time
saw the London-pride (Saxifraga umbrosa) growing
wild.

September 16th. We left Macroom about nine
o'clock, and proceeded to Bantry, a journey of considerable interest and beauty. After leaving
Macroom, the country soon assumed a wilder and
more picturesque character ; and after passing a
village called Inchygeelah, we entered fairly
among the mountains, which we had seen ahead of
us in the first part of the journey. The river Lee,
up the valley of which our road lay, here expands
into a long, winding lake, called Lough Allua ; the
mountains about it, though not high, are finely
broken and varied in form ; a number of bold, rocky
knolls and promontories rising from the margin of
the lake, or jutting into it, give great variety and
richness to the scenery, while the glowing purple of
the heath, mixed with the golden furze, forms a fine
contrast to the grey rocks. As we proceeded, the
mountains rose in higher and bolder masses, and the
whole landscape took a more savage and alpine
character. While our horse was baiting, we walked

to the wild little lake of Gougane Barra, about a 1836 mile off the road, from which the Lee takes its rise. It would be difficult to see a wilder spot than this, or one more striking in its way. The mountains, by which this lake is almost entirely surrounded, are not only higher than those about the Wicklow lake, but much bolder and more varied in their forms, and altogether of a grander character; they are, indeed, remarkably rugged and craggy, and serrated, like the finest of those in North Wales, and so steep that they seem almost to hang over the lake. A single little green island tufted with tall trees, dedicated to St. Finbarry, and considered very holy, forms an agreeable relief and contrast to the general stern and savage character of the scene. We walked up from the lake, by a rough mountain path, to a little farm house, situated in one of the wildest and most sequestered and lonely nooks I ever beheld, beneath enormous, craggy rocks. Neither the farmer nor his family understood English, but they received us very civilly, made us sit down, and offered us each a bowl of goat's milk. The guide who was with us (who himself spoke English with difficulty), told us that this farmer paid £35 a year for his land (in this district they let land not by the acre, but in the lump)—but that the rent was high, and it was as much as he could do to pay it. He had some meadows, and fields of potatoes and oats, at the bottom of the valley, but a large part of his farm consisted of mountain land, not under cultivation. The farm house was different enough, to be

1836 sure, from those we are used to : it was a low, rude
stone building, almost without windows, and filled
with smoke, which made its escape through the door ;
it was divided into two rooms by a partition not
reaching to the roof; the floor was roughly paved
with stones of the most irregular sizes and shapes,
and overhead were the bare rafters.   I cannot
pretend to enumerate all the articles in the kitchen,
especially as the smoke confused my sight, but I
saw a tolerable allowance of crockery, two or three
ricketty chairs, a cradle, and a spinning-wheel.   The
whole affair was more like a Brazilian than an
English farm-house.

The driver of our car, a sharp, intelligent fellow,
told me that the usual rent of cabins was £2 a year ;
that the wages of labour were eightpence a day,
without victuals, and that the poor people would be
very glad to be sure of employment through the year
at that rate, for they were near half the year without
employment.   The guide at Gougane Barra told me
that, at certain seasons, when there was a great
demand for labour, tenpence or even a shilling a day
would be given, but at other times there was no
employment at all, and then they went over to seek,
it in England.

Returning to the car, after our visit to Gougane
Barra, we very soon entered the Glen of Kemineea
a wild romantic rocky pass, in which the driver told
us, there was a " great battle " thirteen or fourteen
years ago,, between the soldiers and the Whiteboys.
Beyond this, the road to Bantry, (a very bad one by

the way) runs along a boggy, moorey valley, and the 1836
country in general is barren and desolate rather than
picturesque. The first view of Bantry Bay, however,
with its inlets and islands and fine mountain boun-
daries, was striking. Bantry itself is picturesquely
situated, but is a wretched little town,—the worst
indeed we have yet seen,—by far the greatest part
of it consisting of poor, filthy, miserable cabins,
as bad as those in the outskirts of Cashel; and the
appearance of the inhabitants is quite in unison with
that of their dwellings. The whole, or almost the
whole, of the town belongs to Lord Bantry, who is
a resident landlord, but seemingly not an improving
one.

September 17th. We started from Bantry for
Glengariff in a very clumsy and ill-contrived car, or
rather cart, with a miserably bad horse, and a man
who knew nothing. The road, after rounding one or
two of the inlets of the bay, ascended to a high,
wild, rugged and moorland country, over which we
travelled to Glengariff. I have seldom, in Europe
at least, seen a worse road than this. The view of
Glengariff, as we descended to it from the height,
was most beautiful,—the smooth lake-like cove
glittering in the sunshine, with its indented rocky
shores, and numerous islands, and the magnificent
peaked and craggy mountains which enclose it. It
struck me as not very unlike the bay of Rio de
Janeiro in miniature, but without the gorgeous
luxuriance of vegetation which gives such peculiar
beauty to Rio. We reached Glengariff inn about

1836 noon, and spent the whole afternoon, till past six, in rambling about the neighbourhood   It is altogether the most beautiful place I have ever yet seen in the British islands.   The cove runs in so deep among the mountains as to have completely the appearance of a lake; the mountains all around it, especially on the western side, are singularly bold and varied in outline, rugged and broken and serrated, and some of them rising into sharp peaks of a more Alpine character than any I saw in Wales or Scotland, and lower down, along the sides of the lake, innumerable knolls and crags and jutting masses of rock, thrown about as it were in the wildest and most fantastic manner, and mixed with heath and small bushes, remind one of the Trosacks. A number of rocky promontories jutting out into the water, and many small islands, on the largest of which is a little fort most picturesquely situated, diversify in the most pleasing manner the glassy surface of the lake (for such it may well be called), while the woods and lawns of Mr. White's place on the eastern shore, and the more distant mountains on the opposite side of Bantry Bay, complete the variety and richness of the scenery.   In all points of view Glengariff is beautiful, but to my taste the first views we caught of it, in descending from the mountains on the Bantry side, were the most beautiful of all.   There is a mixture of softness and grandeur which I have not seen to the same degree in any part of England, Wales or Scotland that I have visited.   It struck me as very singular to see heath,

and moss, and London-pride, growing on the rocks 1836
within a few feet of high water mark—the bottom of
the bank covered with seaweed, and the top with
these land plants. I ought to have mentioned at
Bantry that the sand of the bay, which is composed
almost entirely of broken shells and corallines, is in
high esteem as a manure for bog-land, and a great
deal of it is carried up the country for that purpose.
The great abundance of turf in these parts must be
an important advantage to the poor people.

September 18*th.* Set out from Glengariff in the
same sort of pitiful cart as yesterday, with a
horse that jibbed violently at the first hill
he came to. Edward insisted on returning to
the inn, and taking a fresh horse, with which
we got on rather better. For a mile or two
from Glengariff the road is very beautiful, winding
among picturesque rocks, intermixed with a wild
growth of oak, birch, holly and arbutus, and
commanding the finest views of the surrounding
mountains. One scene in particular, near the road
turning off to Lord Bantry's lodge, is of striking
beauty : the abrupt and craggy mountains here form
a complete amphitheatre, while the deep hollow
which they enclose is diversified with tumbled
masses of rock and tufted wood. From hence the
road soon ascends to the bare, bleak heights, leaving
all richness of vegetation behind. The new road to
Kenmare, which is unfinished, is very good as far as
it goes, but we presently quitted it, and crossed the
mountain ridge by the old road, which is steep and

1836 rough enough in all conscience, but not impassable for a car. The Saxifraga umbrosa, and other nearly allied species of that genus, grow in the greatest profusion on moist rocks by the roadside about the upper part of the pass. After travelling some distance further, among very wild mountain scenery, we descended to the broad estuary of the Kenmare river, about which there is a good deal of wood and cultivation, relieving the savage character of the mountains around. Entering Kenmare we met a funeral procession, which was to me a very new and striking spectacle : there were, I should think, above a hundred people, men and women, the greater part on foot, but some riding, two on one horse; the coffin (which had no black about it) was laid on an open car drawn by two horses, and two old women, enveloped in their dark cloaks, sat bending over it, uttering incessantly a most dismal and monotonous cry. It seemed odd to see not a bit of black about the whole procession.

From Kenmare (a neat, well-built little town) to Killarney, the road is excellent. For the first five miles or so it traverses bleak desolate moors without any beauty, but on reaching the crest of the pass, a grand panorama of mountains burst at once upon our sight,—the long, black serrated ridge of Macgillicuddy's Reeks forming the principal feature, with Carran Tual, the highest mountain in Ireland, standing a little apart from them on the left, and on the right the Purple Mountain and Turk Mountain. The clouded sky was peculiarly favourable to this view, as by throwing the flanks of the mountains 1836 into deep shade, it gave them a striking character of massiveness and sombre grandeur. The first view of the upper lake of Killarney rather disappointed me; from such a height it looked diminutive and poor; but the unfavourable impression was soon removed, and I shall not easily forget the beauty of the drive down to this lake and along its side, through wild woods intermixed with rock and heath, catching at every turn new combinations of mountains. The sun had now broken through the clouds, and shewed the scenery to the greatest advantage, lighting up the lake and the foreground, and throwing the more distant mountains into a kind of hazy purplish shade, while their outlines were shewn distinct and sharp. Macgillicuddy's Reeks, under these circumstances, appeared really magnificent—I had almost said sublime. After this, Turk Lake appeared quite tame in comparison, and the last two or three miles of the road to Killarney are flat and uninteresting.

September 19*th.* We made the tour of the lower and Turk lakes, in a boat, and afterwards landing near the head of Turk lake, walked some way up the Kenmare road, and then back to Killarney. We were very unlucky in having a dull, grey day, highly unfavourable for the scenery. My expectations of the lower lake had not been very high, yet I was a little disappointed—perhaps from seeing it to such disadvantage. Its northern and eastern shores are positively flat and ugly, and though the mountains of Glenà and Tomier, on the opposite side, rise boldly

1836 and steeply from the water, their forms are not fine, and they appear lumpish and heavy, at least on such a day as this. But there are great beauties of detail on several of the islands, and particularly about Glená point : the wood which clothes the shore and fringes the broken rocks along the water's edge, though not of lofty growth, is very rich and beautiful, with a remarkable variety of foliage and colouring; the dark, glossy green of the holly, and the more sombre tint of the yew, and the bright, deep verdure and delicate flowers of the arbutus, are finely mixed with the lighter greens of the deciduous trees, and with the yellow and reddish autumnal tints which are already appearing in several places.

Inisfallen disappointed me : it is low and flat, and by no means thickly wooded, though the individual trees are fine, especially the hollies and yews, the former of which grow to a much larger size than I have ever seen elsewhere. The remains of the abbey (which is said to be very ancient) are neither extensive nor beautiful. O'Sullivan's Cascade is a pretty little toy of a fall, very well shaded and sheltered by thick wood and mossy rocks, but the height is inconsiderable, and the stream very small. But if the lower lake on the whole disappointed me, the channel between that and Turk Lake made full amends. It would be difficult to find a more lovely scene in its way than the smooth, still, glassy water, dark yet clear as crystal, reflecting with exquisite distinctness the luxuriant thickets and tall trees which border and overhang it on both sides, while

peeps are caught at intervals of the craggy Croma- 1836
glown mountain and the bold Eagle's nest.
Dinis Island, which forms one side of this channel,
is very richly wooded, and the arbutus trees upon it
are quite magnificent, – not shrubs, but really great
trees, of a very picturesque character, with thick,
sturdy, crooked and twisted trunks, and spreading
heads. I never before saw the arbutus growing to
half the size it does here. The banks of Turk lake
are finely wooded, but its form is too round and
regular, and its surface not diversified by islands,
and the mountains around it rather deficient in
variety and boldness of outline; though it must be
owned that Turk mountain, in some points of view,
is a fine rugged fellow; as seen from the lower lake,
and from near Killarney, on the contrary, he appears
round and lumpish. Turk waterfall I saw late in
the afternoon, and in a very hurried way, but it
seems to be much higher and finer and more
picturesque than O'Sullivan's, though like that,
sadly deficient in water at this season.

September 20th. We made an excursion to the Gap
of Dunloe, a deep rocky cleft or gorge separating the
underfalls of Macgillicuddy's Reeks from Tomies
and the Purple Mountains. It is a fine specimen
of rude and savage mountain scenery, gloomy and
desolate, and with a grandeur of effect quite dis-
proportioned to the real magnitude of its dimensions.
The mountains rise very abruptly, all but precipitously, on both sides, towering to a great height,
cliff above cliff and ledge above ledge, black and

1836 rugged and bare, with no vegetation but moss and a
little heath in the interstices and on the shelves of
the cliffs; while enormous broken masses of rock
tumbled about in the wildest confusion, encumber
the bases of the mountains, and a chain of little
black lakes, occupying the bottom of the gorge, add
to the wild and singular effect of the scene. I
walked through the pass, and far enough to get a
view of the upper lake,—but it certainly does not
appear to advantage from a height.

The Saxifraga umbrosa is exceedingly abundant in
the Gap of Dunloe, ornamenting all the clefts and
interstices of the rocks with thick tufts of its
beautifully neat foliage. It is scarcely less plentiful
on shaded, mossy rocks in the woods about Turk
and the upper lake, where it grows larger, and
assumes a slightly different appearance. S. Geum
is far less common, and grows, as far as I have seen,
only on very wet rocks, where there is a drip of
water; I found it by the roadside on the banks of
Turk lake, as well as in the Gap of Dunloe.

September 21*st*. We spent a very pleasant day
going in a boat the whole day to the head of the
lake and back again. The upper lake is very
different from the other two, and to my taste far
more beautiful: it is quite surrounded by mountains,
and its surface is studded with a multitude of bold,
rocky islands, tufted with heath, and furze, and fern,
and with the richest thickets of oak and ash and
holly and arbutus and mountain-ash. The varied
forms and colouring of these islands, with the glassy

surface of the water reflecting them like a mirror, 1836
the woods that clothe one side of the lake,
the mixture of heath and broken rock on the
mountain sides above them, and the magnificent
mountain outline at the head of the lake,
form a most exquisite combination of scenery.
Even Glengariff falls short of it. Macgillicuddy's
Reeks, which rise most conspicuously in view as one
goes up the lake, with their abrupt black sides and
sharp jagged outline, have an effect of height and
grandeur beyond any other mountains of their class
that I have ever seen.

Besides the plants I have mentioned, Osmunda
regalis abounds on the islands and shores of the
lakes, especially on Dinis Island. The beautiful
Hymenophyllum Tunbrigense grows in the utmost
profusion among moss on damp shaded rocks, in the
woods and near the waterfalls. H. Wilsoni grows
in the same situations, and often mixed with it, but
is less plentiful. The variety of mosses, Junger-
manniæ, and Lichens, is very great, but it is now
too late in the year for the generality of flowering
plants.

September 26th. From Killarney to Tralee we
passed over a very dreary open country, a great
proportion of it bog, without picturesqueness or
fertility, or beauty of any kind ; and I have never
seen more miserable cabins than in this tract. One
can hardly believe them to be intended for human
habitation. Such as they are, however, there are a
great many of them, and the numbers of ragged

o

1836 half-naked children that one sees about almost every
cabin, show that the population is increasing rapidly.
Both men and women are generally in rags, and
have a haggard and poverty-stricken look, and the
children frequently have not clothing enough for
decency, to say nothing of comfort. The pigs alone
look sleek and comfortable. At a poor village called
Castle Island, where we changed horses, the coach
was surrounded by a terrible swarm of beggars, who,
I must say, all looked really in want of assistance.
Tralee is rather a large and well-built town, with an
air of considerable business and prosperity, and
evidently improving : it has several good streets, and
many respectable shops, among which, however, I
did not observe a single bookseller's. The only
public building at all remarkable is the court-house,
which is classical and handsome. There is nothing
of interest either in the town or its neighbourhood,
so that we passed a dull afternoon here ; but after
dinner I was rather interested by a conversation with
two of our fellow travellers, who were Catholics and
red-hot repealers ; one of them declared that for his
own part he wished to see Ireland entirely independent of England, and that he believed she would be
able to assert and maintain her independence against
any power whatever. He seemed to me to under-
rate very much the strength and importance of the
Protestant party in Ireland. The other, although a
priest, was apparently more moderate, and said that
the Catholics desired nothing more than complete
political and civil equality—or in other words, to be

on an equality in the eye of the law with the 1836
Protestants,—they did not desire ascendency. Both
however admitted that the condition of the country
on the whole was improving, that an unusual degree
of tranquillity now prevailed throughout it, and that
a better prospect altogether seemed to be opening
for Ireland.

September 27th. We travelled by the mail from
Tralee to Limerick. I saw many trunks, or portions
of the trunks and roots of trees, which had been dug
out of the bogs. At last, approaching Newcastle,
we descended from the moory uplands into the
extensive and rich plain of Limerick, which stretched
out before us as far as the eye could reach, and
presented a very agreeable contrast to the desolate
tract of country we had just crossed. All the rest of
our day's journey was over this plain, which is more
level than any other part of Ireland I have seen, and
apparently still more rich and fertile than Tipperary
county, but the deficiency of trees and hedges
deprives it of all beauty. A great part of the land is
in meadow, the rest produces wheat, barley, oats,
potatoes, and flax. There was also in general
(though with some exceptions) a decided improvement in the appearance of the cabins, which were
larger than the common run of those in Kerry, and
had chimneys and windows—small indeed, but still
glazed windows. It struck me too that there was
an improvement in the looks of the women, and at
Rathkeal in particular I noticed one very pretty girl
at the door of a cabin. Rathkeal, which I believe is

1836 the second town in Limerick county, is a tolerably
good-looking town, and of some considerable size ;
but here, as at all the other places where we changed
horses, the swarms of beggars that besieged the
coach were quite distressing.   I was struck not only
with their number and perseverance, but with the
variety of the solicitations, and I might almost say of
the arguments which they used to gain their point.
The ruins of several ancient buildings, and a broad
though shallow stream, and the groves of Lord
Dunraven's demesne, and fine tall trees shading the
road, give considerable beauty to the neighbourhood
of Adare.   There is nothing striking in the approach
to Limerick, but the broad handsome street down
which we drove to the hotel gave me a favourable
impression of the city.

September 28*th*.   Rambling about Limerick this
morning, we fell in with Mr. W. O'Brien, whom I
had known in London, and who kindly undertook to
be our guide to everything worth seeing, and was of
great use to us. He shewed us the House of Industry,
the County Infirmary, the Lunatic Asylum, and the
County Gaol. But in visiting these institutions, I
felt strongly my deficiency of knowledge with respect
to those of the like kind in England.

The House of Industry, I confess, struck me as
not being well managed ; it contains four distinct
classes of inmates, who ought surely to be kept
separate,—infirm or aged and helpless poor—lunatics
and idiots—disorderly women and vagrants—and
young children. The want of any effectual separa-

tion between these seems very objectionable. But I 1836
believe there is a want of means, which may serve to
excuse this defect; and indeed the scale of the whole
establishment is quite inadequate to the wants of
such a population as that of Limerick. It contains
about 470 inmates of all descriptions, among whom
are 52 girls and 62 boys; the schools for these seemed
to me the best managed part of the establishment.
The rooms in which the infirm and old were lodged
struck me as being close and ill-ventilated. This
establishment is supported partly by Grand Jury
presentments, and partly by voluntary contributions.
The Infirmary seems to be well managed, but what
struck me most here was the number of patients
with fractured skulls and other severe injuries received in frays. By their own account, however,
none of these men were given to quarrelling,—they
had all been assaulted without provocation. The
person who attended us about this institution said
that fractures of the skull from such causes had
become much more frequent of late years, owing to
the use of stones instead of sticks in the faction fights.
The Infirmary is maintained partly by presentments
and parliamentary grants, partly by subscriptions,
partly by fines; but its funds appear to be sadly
deficient. I was very much struck with the extreme
cleanliness, quiet, and good order of the Lunatic
Asylum. As far as I am able to judge, nothing can
be better managed. This is supported entirely by
Grand Jury presentments. The Gaol also appears
to be under excellent management. By the way,

1836 the gaoler told us that Lord Mulgrave, when he visited this gaol, had released one prisoner—an unfortunate woman whose husband was absent, and whose numerous children were starving in consequence of her imprisonment. So much for the outcry raised by the Tory papers against Lord Mulgrave as setting all the rogues and vagabonds at liberty.

Limerick has nothing picturesque in its situation, but it is a handsome city, at least the new part of it; George Street and some of the other principal streets are full as good as any in Cork, and perhaps of a more aristocratic appearance,—and I think one sees a greater proportion of smart and gay looking people than in that city. The shops seem to be very good and among them are several respectable booksellers. I may here mention what has more than once struck me as a sign of the country,—this is the great number of theological and controversial works that one sees in the bookseller's shops in all the great cities of Ireland. The old town of Limerick is very different indeed from the new, and its population has an extremely poor, squalid, tattered look; but I believe that what one sees in the streets gives a very imperfect idea of the real poverty that exists. The first view of the Shannon disappointed me ; it is not here a handsome river, nor is it by any means so large as I expected. The old bridge—Thomond Bridge—which is said to have been built about 1210, has been pulled down, and a new one is building in its place.

September 29th. We made an excursion to the 1836 village of Castle Connell, six miles above Limerick, in order to see the rapids of the Shannon. Having reached the village, we took a boat, went down some little distance, landed in the grounds of Sir Hugh Massey, just above where the principal rapids begin, and walked along the river-side so as to get a full view of them. They are certainly very beautiful, and very peculiar; I never before saw rapids on a large scale. They reminded me much of Humboldt's description of the cataracts of the Orinoco. The river, about 200 yards wide, and beautifully clear, rushes with great swiftness over a long succession of ledges and barriers of rock, leaping and curling and foaming and whirling round and round in innumerable eddies. In some places it is all in a foam from bank to bank; in others, long shelves of rock covered with black moss, lift themselves above the surface, and in the intervals between them the dark transparent water hurries swiftly down the slope; in others it is broken by rocks and stones which it just covers, and over which its white waves go leaping and dancing in wild tumult and confusion. The fall is nowhere considerable, but the rapids continue for a long way together, and as far as mere water (without accessories) is con_ cerned, I have seen nothing half so beautiful in Ireland. The banks are flat, but their bright verdure and groves of tall trees and numerous gentlemen's seats, give them a certain degree of quiet beauty.

October 7th. We have been staying all this week at Dromoland, the seat of Sir Edward O'Brien, fourteen

1836 miles from Limerick and five from Ennis. The house, which is still unfinished, is one of the noblest specimens of modern castellated architecture that I have ever seen, and the grounds are extensive and pretty, without anything striking. There is a small natural lake just in front of the castle, and much of the timber about the place is fine, but not to be compared to that of Woodstock or Curraghmore. The surrounding country, which consists entirely of limestone (the carboniferous or mountain limestone) is undulating, and in some parts rising into considerable hills, but is much too open and bare for beauty. The greater part of it is enclosed and cultivated, either as pasture or arable, though there is some bog; and the condition of the people generally appears to be pretty good. Sir Edward resides constantly on his own estate and takes care to provide employment for the peasantry, and his family, who are very religious, pay great attention to the condition and wants of the lower classes, and I believe do a great deal of good. I have seen hereabouts scarcely any of those miserable cabins which in Kerry and Cork meet the eye at every turn; the cottages are of a substantial and decent appearance, well thatched, with good windows and chimneys, and a fair supply of furniture. Still, however, one often sees tokens of that slovenliness, that disregard of neatness and comfort, with which one is perpetually struck in the South of Ireland. The walls of the cottage are often black with smoke; the floor and furniture are dirty; a broken pane in the window is filled up with an old hat or an apron; the pig and 1836 the poultry frequently share the apartment with its human inmates; and the tenant himself generally wears ragged clothes, while an Englishman in the same circumstances would wear whole ones. I am beginning to see more and more clearly that the real condition of an Irishman is not to be judged of by outward appearances. Lady O'Brien has told me that there are frequent instances of men living apparently in the utmost poverty, and leaving large sums of gold to provide for the expenses of their funeral, and to have masses said for their souls. I am, however, very glad to learn from my friends here that there has been for some years past a real and evident and progressive improvement in the circumstances and comfort of the lower classes in this country, though there is still abundant room for further improvement. The farms here, as in all other parts of Ireland that I have seen, are generally small, though there are some men who hold as much as one hundred Irish acres. The usual wages of labour are tenpence a day. Sir Edward affords employment to multitudes of people, not only in agriculture, but by the alterations and improvements going on about his house and grounds, and by the attempts he is now making to discover mines.

I have learned from the O'Briens several particulars relating to the "Terry Alt' disturbances in this country five or six years ago. A most formidable combination existed among the peasantry, so well organised that for a good while the magistrates and police and military were completely baffled, and no

1836 evidence could be procured against any of the criminals. Bands of armed men assembled in the night, dug up the pastures or carried off the crops, and compelled those who were more peaceably disposed to join them and take their secret oaths, under pain of death ; the houses of the farmers and small gentry were broken into and searched for arms; threatening notices were put up on the doors of those who had done anything displeasing to the insurgents, and many were murdered, sometimes two or three in a night, without any clue to the murderers. The county was in a state of utter lawlessness and insubordination. At length, from some cause or other, one of the criminals gave information which led to several arrests ; and then, as if the spell were broken, many others came in and gave evidence. A special commission was sent down to try the prisoners ; a number of the neighbouring gentry attended on the juries; many of the prisoners were convicted, sentenced to transportation, and sent off immediately from the court-house to the hulks, without being allowed to see any of their friends. The effect of these vigorous measures is described as being strikingly rapid ; order and tranquillity were restored almost before the Commission had terminated,—the illegal combination was effectually broken, and the country has been quiet ever since. Perhaps the same effects might have been produced in other disturbed districts by similar means, without the necessity of a coercion bill. The people, it is said, imagined that the Government was secretly favourable to their proceedings, till they were undeceived by the special 1836 commission.

Sir Edward O'Brien told me, that at the time of these disturbances, the Catholic priests, who preached against the outrages and exerted themselves to dissuade the people from them; lost all their influence and were themselves insulted and threatened.

LETTERS.

Cratloe, near Limerick.

October 21st, 1836.

My Dear Father,

I am really ashamed of having so long put off writing to you and acknowledging your letter, which I received at Killarney. I never was more shocked than at hearing of the very unexpected death of poor Caroline* ; it was one of those sudden and unlooked for blows which one can hardly believe possible, for no one in this world seemed to have a fairer promise of a long and happy life. It makes my heart bleed to think of poor Henry and his children. I am afraid it must also have been a very severe shock to dear Emily. We have now been about three weeks in this neighbourhood, staying at different houses—first at Dromoland Castle, Sir Edward O'Brien's—then

* The wife of Captain Henry Napier, Lady Bunbury's brother; she was very beautiful and died of Cholera at a villa near Florence, leaving four young children.

1836 here with our friend Stafford O'Brien*—next with
Mr. Spring Rice† at Mount Trenchard—then at
Lord Dunraven's at Adare—and now here again;
and the time has passed very pleasantly. Of all
these new acquaintances, those that I like best are
Mr. Spring Rice and his family. He was very kind
and expressed great pleasure at becoming acquainted
with us, and I had a great deal of conversation with
him about Irish History and politics, the state of
the country, literature, etc., from which I derived
much pleasure, and I hope, profit. He seems to me
the most sensible, moderate, and clear-sighted man
I have met with in Ireland, and thoroughly
acquainted with the country and the people, so that
one learns a great deal from him. Miss Spring Rice‡
is an uncommonly pleasant girl. The O'Briens of
Dromoland are in many respects a pleasant family.
Our friend here, whom I used to think the most ultra
of Tories, appears quite moderate in comparison
with others I have met in Ireland. He is doing a
great deal of good on this estate, which his father has
given up to him, and Tory as he is, he is quite
enthusiastic in his praises of the peasantry. I have
been very much pleased to learn from everybody I
have conversed with in this part of the country, that
the condition of the people generally is visibly and
progressively improving for many years past, though

Afterwards known as Augustus Stafford.
† Afterwards Lord Monteagle.
‡ Maid of Honour to Queen Victoria, afterwards married Mr. James
Marshall of Leeds.

everybody admits that there is still much room for 1836 improvement. Certainly the apparent condition of the people is strikingly better in this county (Clare), and even in Limerick, than it is in Kerry and Cork, but it varies very much on different estates, and one may discover almost at a glance, whether one is on the property of a resident and wealthy and improving landlord, or otherwise. At the same time I am convinced that one is liable to fall into great errors and exaggerations by judging too absolutely of the condition of the peasantry from external appearances, according to our English ideas; for the Irish evidently do not care about neatness and comfort in the way the English do; and moreover, the difference of wages is not nearly so great in reality as it sounds. I think there is a strong tendency to exaggeration in Inglis' book, not perhaps actual mis-statements, but a disposition to look at the dark side of things, and to make the worst of everything. It is however, too true, I believe, that there is a great deal of destitution and misery in the towns, and above all, I am told, in Limerick; and I have never heard any valid objections to a poor law for the support of the old and infirm and sick, and widows, and helpless orphans.

There is no striking beauty in this part of the country, from Limerick downwards, but it is a fertile and pleasant land, and numerous ruins of abbeys and other old buildings give it a degree of interest, and the Shannon itself is certainly a very fine river. The rapids of the Shannon at Castle Connell, six miles

1837 above Limerick, are very beantiful and very peculiar ;
they reminded me of Humboldt's description of the
cataracts of the Orinoco.   I was delighted with the
scenery of Killarny, especially with the upper lake,
which is certainly more beautiful than anything I saw
in Wales or Scotland, though I cannot agree with those
who prefer it to the Swiss or Italian lakes.   The outlines of the mountains are finer even than in Merioneth-
shire, and the richness and variety of the foliage on
the numerous islands—the mixture of oak, and ash,
and holly, and arbutus, and mountain-ash—give a
peculiar charm to the scenery.   Glengariff is not far
inferior in beauty to Killarney.

      We leave this to-morrow for Dublin by way of
Killaloe, Portumna, and Banagher, and shall probably stay a week at Dublin, where I hope to hear
from you or Emily.   I shall be very glad to get
home, and spend the winter quietly with you.

                  Believe me ever,
                        Your very affectionate son,
                              C. J. F. Bunbury.

                        ———

                        JOURNAL.

January 21*st.*   *Barton.*   My reading has been
interrupted for the last week by the Influenza.   I
will take this opportunity of reviewing what I have
done, and in some measure, what I have thought,

since my return from Ireland, rather more than two 1837 months ago. I cannot boast of my diligence in reading or writing. I have read O'Driscol's "History of Ireland"; the two pamphlets by a "Manchester Manufacturer", which in quantity of matter are equal to two ordinary octavo volumes; the first volume of Smith's ".Wealth of Nations"; the greater part of the Evidence repect-ting the Irish Poor; a considerable portion of Napoleon's Memoirs; the fifth volume of Napier's "Peninsular War"; Whewell's, Buckland's, and Bell's Bridgewater Treatises;—besides some odds and ends. But I have certainly studied history and politics less diligently than I did last winter in London, having had many distractions, and having yielded a good deal to my natural predilection for natural science, which would certainly be my favourite and principal pursuit, were it not for my father's anxious wish that I should try to distinguish myself in public life. My fits of ambition have been feeble and short-lived, notwithstanding the encouraging success of my speech to the electors of Bury, on the fifth of December. That success gratified me much at the time, and kindled a degree of ambition and hope of further successes, but the effect was not lasting, and my exertions soon relaxed. It is not mere indolence that causes this; I feel a secret consciousness that I have not, and never can have, that stern, determined, unbending vigour of character, which is necessary to success in troubled and revolutionary times; I am too mild, too yielding, and at the same time too

1837 scrupulous and fastidious, for a public station in such times. Nature certainly intended me for quiet times and a private station. I have sometimes wished that I was not an eldest son.

But as I am placed in the station of a country gentleman, I must make the best of it, and exert myself to fulfil the duties of it with honour. Since my return to Barton, I have begun acting as a magistrate, and have got in some degree initiated into that kind of business, which is not very difficult, if it does not offer any particular interest.

February 1st. There is a good article on the New Poor Law in the January number of the " London and Westminster Review ". The writer signs himself W. E. H. He exposes the numerous fallacies which have been propagated on the subject, partly by factious and ill-disposed Tories or Ultra-Radicals partly by benevolent and well-meaning, but ill-judging, people. He shows :—

1. That the new law, which by its enemies of both parties is imputed to the Whigs, is as much the measure of the Radicals as of the Whigs ; that it was not only warmly supported, but in great part planned and drawn up, by staunch Radicals ; and that the germ of it is to be found in the writings of Benjamin Franklin.

2. That the cases of hardship and oppression which have been brought as grounds of charge against the Bill, have almost all been found, on careful enquiry, to be either false or exaggerated.

3. That the principle of the Bill is to give relief to

the destitute, but not on such terms as to make the 1837 condition of the pauper better than that of the independent labourer.

4. That the strict discipline of the Workhouse is the only effectual means of attaining this end.

5. That the food and lodging in the new work-houses are generally better than in those under the old system, and in some parts of England even better than what the labourer enjoys in his own cottage.

6. That those very parishes which, under the old system, were guilty of the worst neglect of the old and afflicted, have been the loudest in their denunciation against the cruelty of the new law.

7. That all the money which is saved from the poor-rates, by an improved system, is so much added to the regular and proper fund for the employment of labour; and that, in fact, the labouring classes have been in unusually good employment since the new law came into operation.

8. That in the actual situation and circumstances of England, the really industrious and prudent can seldom fall into extreme distress, and that the operation of the improved system of poor laws will necessarily tend to render industry and prudence more general.

February 3rd. Yesterday's paper contained the debate on the address in both houses. The measures to which the attention of Parliament is thus specially called are many and important:—Poor Laws for Ireland, the English Church, Irish Tithes

P

1837 and Corporations, Law Reform, the affairs of Lower Canada and some others. The mover and seconder of the address in the House of Lords were Lord Fingall and Lord Suffield,—in the House of Commons Mr. Sanford* and Mr. Villiers Stuart. There was no amendment to the address in either House, but the Duke of Wellington and Sir R. Peel both found fault with the policy of Ministers in regard to Spain. The Duke also objected to what Lord Fingall said of the increased tranquillity of Ireland. Mr. Roebuck made a violent ultra-Radical (or rather Jacobin) speech, furiously attacking not only the Ministry of the Whigs in general, but O'Connell and the more moderate section of the Radicals also; he was loudly cheered by the Tories, who were of course delighted at this manifestation of a schism in the Reform party.

If he is in his senses, I can hardly imagine what his object can be, unless to make a revolution, and to be himself the Robespierre of it. Mr. Hume and Dr. Bowring opposed his motions, and declared their intention of supporting the present Ministers.

February 5th. On reading a fuller account of Mr. Roebuck's speech, in the *Speetator,* I find that my first impression exaggerated the violence of it. In fact, it is not so remarkable for intemperate

---

* Edward Ayshford Sanford of Nynehead Court, Somerset, was in Parliament from 1830 to 1840; was in five Parliaments: two for the whole county of Somerset, in three for half the county after the Reform Bill, and always supported the Whig Ministry. He was at one time proposed by a section of the party, the more Radical section, as a Candidate for the Speakership, but he did not allow his name to be publicly proposed as it would have embarrassed the Ministry. He was an active member of many important Committees.—F. J. B.

language, as for conceit and arrogance, and for the 1837 extreme intolerance and unfairness of treating the Whigs as deceivers and hypocrites, because their opinions do not go the whole length of their own. He says to them, in substance. — You pretend to be Reformers, but you will not carry your reforms to the length of totally overturning the Constitution, and establishing a naked unmixed democracy; *therefore* you are totally false and insincere in your professions. This is just the same sort of bigotry, on a different subject, which has so often been displayed by churchmen. Besides, what possible right or claim has Mr. Roebuck to set himself up, as he does in this speech, as *the representative* of the people of England ?

February 6th. The Constituent Assembly of France contained by far too great a number of mere theorists and projectors, and too few practical men, or men of business; this was no doubt one main cause of the many errors and imprudences it committed. I am inclined to think that the very reverse of this is rather to be apprehended and guarded against in our reformed Parliament, especially when its constitution shall have been rendered still more democratic by the enactment of triennial Parliaments and the Ballot. It is very desirable that in every legislative and deliberative assembly there should be a certain number of theoretical and speculative men; there may easily be too many, but a certain proportion there should be, if it be desirable, as it surely is, that legislation should be con-

1837 ducted on sound principles, and with large and enlightened views. In matters of detail, and in the adaptation of particular provisions to the peculiar circumstances of this or that class or district, they are likely to be inferior to men of business, but on the other hand, they are used to considering a subject in its whole, and not merely in parts; they are used to looking at a question in all possible points of view, not only by itself, but in its connection with other questions, and in its bearings upon other times and countries. Their habits of mind lead them to generalize, to combine, to compare, to follow out certain general principles which they have established, and to try any proposed measure by the test of those principles. Hence they are very useful in correcting the narrow partial notions and one-sided views of mere practical men. They may not be free from prejudices,—few men are,—but their prejudices will be entirely of a different kind from those of practical men. In a country like England, essentially a country of *business*, there is not likely to be any deficiency of practical men,—men of business,—men who have been used all their lives to matters of business and detail. Valuable and indispensable as such men are in the conduct of affairs, they yet require, as it seems to me, to be mixed and tempered with a due proportion of philosophical and theoretical men, in order to form a really good legislative body. Consequently, any law of election which would tend to exclude this latter class from Parliament, is undesirable.

February 7th. I can very easily conceive the 1836 feelings with which a sincere and conscientious Tory may regard the monarchy, the aristocracy, the established church, and other old institutions of the country. He may look upon them in the same sort of light as upon one of our venerable cathedrals or our noble though ruined abbeys ; they may be hallowed in his eyes by antiquity and by the glories associated with them, as well as by the impressions of education, and by a fancied connection with the important truths of religion. Thus he may come to regard those institutions with a mixed feeling of romantic or poetical enthusiasm and of religious veneration, till their very defects and abuses appear to him to be beauties ; or, if he is sensible of those abuses, he may yet be unwilling to meddle with them, lest the whole edifice, so sacred in his eyes, be endangered by the removal of any one particle. Such, I apprehend, are the sentiments of many High Tories, who are among the most upright and honourable of men—of my friend Stafford O'Brien for example. But it strikes me that neither the most active Tory leaders in Parliament, such as Sir Robert Peel and Lord Lyndhurst, nor the principal Tory journalists, ever appeal to sentiments or principles of this kind. They take lower ground, and dwell mostly on considerations of expediency and utility, except when they choose to have recourse to the argument (which they always misapply) of the sacredness of property, and to the sophism of confounding trusts with rights.

1837   *February* 10th.   The papers of yesterday and the day before are full of the two nights' debate (on the 7th and 8th) upon the introduction of the Irish Municipal Reform Bill.   Lord John Russell made an excellent speech on bringing it in, and took occasion to enlarge upon the whole subject of Ireland, to defend the policy of Ministers and the conduct of the Lord Lieutenant, to shew the decrease of crime and disturbance in that country, and to lash with deserved severity the resolutions of the (so-called) Protestant meeting in Dublin.   It appears by his speech that the Ministers have very properly determined to stake their ministerial existence on the success of their measures of justice for Ireland, and to resign if unable to carry them.   O'Connell did not speak well; he got into a passion at the attacks made on him and his friends by the Orange Sergeant Jackson, and, neglecting the real subject of discussion, ran into coarse and angry personalities.   Lord Morpeth, on the 2nd night, made a clear, temperate and satisfactory defence of the Irish administration.   Sir James Graham and Sir Robert Peel were the principal speakers on the other side, and both spoke well, but without (in my opinion) at all confuting the arguments in favour of the measure, or shaking the ground on which they were brought forward.   Sir James Graham dwelt much on the power and mischievous designs of the Irish association and the danger to the Irish Church which would be the consequence of extending municipal rights to the Catholics.   This in fact is the only thing properly to be called an *argument* that is urged against the 1837 measure by any of the Tories. Peel's speech strikes me as very dexterous and artful, admirable fencing, but nothing more. The bill was at last allowed to be brought in without a division.

February 16*th*. Arrived in London the day before yesterday. I heard to-day that Lord De Roos, instead of going abroad again, as I had of course expected from the event of the trial, means to stay and brave public opinion,—nay, what is more strange, that he will be countenanced and supported by many in the fashionable world. If this be true, it betokens a queer notion of honour and principle in the highest classes. Is it because of Lord De Roos's uncommon power of pleasing in society, that his habit of cheating at cards is to be overlooked—or because there are many among his equals who have a con- sciousness of not being much better than himself? Certainly, though the evidence was conclusive against *him*, it shewed several of the witnesses in a light not much more favourable, who as persons who, if they did not positively cheat, yet made gambling their principal resource, and actually depended on it for their subsistence. Gambling in any way is odious, but most particularly so when it is made a profession and a source of income.

February 20*th*. Spent some time at the British Museum. The Egyptian paintings are very curious, —the colours surprisingly well preserved ;—scarcely any shading or perspective. The artists of that nation were much more successful with animals than

1836 with man; the geese, ducks, herons, and other wild-fowl, and the oxen and even horses, are painted with much spirit and accuracy, while the figures are exceedingly stiff, awkward, and ill-proportioned. These figures are all painted of a deep red-brown colour, with black hair and eyes; the people of condition are clad in long garments, while the slaves or inferiors, both men and women, are nearly naked. In some of these paintings, the ladies are represented sitting on chairs very like those now in use, and smelling the flowers of the white lotus or water-lily. In others the common people are prostrating themselves and kissing the ground before their superiors.   In the mineral room, among other things I noticed a very large crystal of Titanite from Minas Geraes, like those which were given me by the Baron of Catas Altas; also Titanite in a delicate capillary or mossy form, of a bright copper-colour, from Moustiers in Savoy; a fine stalactitic specimen of brown hæmatite (I presume from Brazil) with all its interstices filled with cupreous arseniate of iron. The series of phosphate of lead is extremely beautiful, exhibiting almost every shade of green, brown, yellow and orange; the arseniates of copper are of striking beauty, but of the cupreous arseniate of iron I do not think there is any specimen so fine as those which I procured in Brazil.

LETTERS.

MY DEAR EMILY,

I am sorry to hear from Miss Fox that my father
has been ill again and that you have put off coming
to town. Mrs. Gwyn is sadly vexed at it, for she
had been calculating too confidently on his coming,
and I do believe that the sight of him would do more
than anything to make her well again. She is
looking very weak and very much pulled down, and
is terribly nervous about herself. Besides them I
have seen Miss Fox, the Somervilles, Mrs. Udney,
Mr. Broadhurst, Mr. Jones, and divers Cantabs
whom you do not know. Mrs. Broadhurst has been
very ill, and though she is now better, she was not
able to see me, but I had a long and pleasant
political talk with Mr. Broadhurst, who, I found to
my surprise, had read my Bury speech. Dear Miss
Fox seems very well, and very glad to see me, and
I had a long comfortable chat with her, and while I
was there, in came the Archbishop of Dublin* and
his wife, and so I had the opportunity of becoming
acquainted with them: he is certainly a very odd

* Archbishop Whateley.

1837 uncouth looking man, but very entertaining, a great
talker, and seemingly quite ready to launch out on
any subject. Upon that occasion he rambled from
toads and frogs to Sydney Smith and the Bishops!
Miss Fox says, I think very justly, that there is too
much purple and fine linen in Sydney Smith's
pamphlet,—too much love of wealth ; that it is quite
conclusive *against* the Bishops, but not for the
Chapters, and that the result will be, that everybody
will believe the Bishops against the Chapters and
the Chapters against the Bishops. I can hear
nothing positive about a dissolution,—everybody
seems uncertain whether there will be one or not.

I have not caught even a cold, but am very well
and very thriving, extremely well amused, and should
enjoy myself mightily if it were not for the horrid
weather. I saw Hanmer for a minute on the road,
I hope he will soon come to town; and you too, the
more as I am sure that all the doctors in the world
could not do Mrs Gwyn so much good as my father
could. Pray give my best love to him and Hanmer,
and believe me,

Yours very affectionately,

C. J. F. BUNBURY.

1837

JOURNAL

February 24*th*. The grand division on the Irish
Corporation Bill took place on Wednesday last, after
a debate of three nights. Lord Francis Egerton
was again this year, as last year, the ostensible
leader of the opposition, having moved on Monday
that the corporations should be abolished ; but his
motion was rejected by a majority of 80 (322 to 242)
which was a much better majority than I at least
had expected. Last year it was only 64. The Tory
newspapers seem to be confounded and perplexed,
and declare that this division is unaccountable. I
suppose now we shall have no dissolution, at least if
the present ministers continue in office, though Miss
Fox seems to think they will resign in case the House
of Lords should reject this measure. Some of the
speeches, on both sides, were very good,—Mr.
Charles Buller, Lord Howick, Mr. Roebuck, and Shiel,
on the Liberal side,—Peel and Sir James Graham
on the other. But a great deal too much of
the debate turned on the merits of Lord Mulgrave's
administration, which did not strictly belong to the
question.

February 27*th*. It is amusing to see the dismay
and perplexity of the Tory newspapers at the great
division on Wednesday, and the feeble attempts they

1837 make to console themselves, or their readers. Still in spite of the increased majority in the House of Commons, I am afraid there is no chance of this most important measure passing the other house; on the contrary, the High Tory party (of which the focus is in the House of Lords) shew a more obstinate and bitter spirit than ever; and as they are well aware of their superior strength in that house, they will in all probability use it unsparingly. Then the Ministry will probably resign, and either Peel or Lyndhurst will be commissioned to form a new one. But it is evident enough that they cannot go on with the present House of Commons; will they venture to appeal to the country? and will they by so doing gain a majority? This I doubt extremely. In short I do not see how we are to get out of this awkward *fix*, unless the Lords should be suddenly enlightened by special grace. As matters now stand, a Liberal ministry cannot manage the House of Lords, nor a Tory ministry the House of Commons. What will be the end? God preserve us from a revolution. Hanmer tells me that the bad policy of the present government of Greece causes all the most fertile lands in that country to remain waste and desert. These lands, having all been monopolized by the Turks, on *their* expulsion came into the hands of the actual government, which will neither give them away to individuals, nor sell them cheap, nor even let them at a low rent; and the people are all too poor to take them at the rent demanded; consequently they remain uncultivated and useless. The country, he says, is in a most wretched state of 1837 poverty, and many of the people are actually emigrating to the Turkish provinces, finding they can live better under the Mahometans than under their own government.

February 28*th*. I hear that the majority of 80 on the Irish Corporation Bill was a surprise to the Ministers themselves, as well as to their enemies. I hear too that they decidedly mean to resign if the Lords should, as is probable, adopt the amendment which was rejected by the Commons. On the other hand, I hear it is the opinion of some politicians (in the city I presume) that the Lords will not adopt so bold a measure this year, but will content themselves with cutting and hacking at the details of the Bill, as they did in the case of the English Corporations. If they do no more harm than they did to that Bill, we may be well enough satisfied; but it seems hardly probable that they will give way thus, after their boasting and all their invectives.

March 1*st*. Horribly cold. I went to the Linnean Society, and looked at Aublet's Plants of Guiana, at Humboldt's Nova Genera, and at some fasciculi of the Flora Graeca, the plates of which (by Ferdinand Bauer) are remarkably beautiful. In Aublet I recognized some plants that I had found in Brazil, but not many; his plates, which are uncoloured, are not highly finished, nor very delicately executed, but in general very characteristic, and accompanied with useful though not very elaborate dissections. The

1837 names he has given to his new genera are mostly very barbarous.

March 2nd. At the Linnean Society I looked through the first volume of Dr. Wallich's splendid work, the Plantæ Asiaticæ Rariores, comprising 100 new plants from the Himalaya Mountains and the Burmese empire, many of them extremely interesting. The most striking of all is the Amherstia nobilis, a tree nearly related to the Bauhinias, and perhaps the most splendid of known plants, with its long pinnated leaves, and its large pendulous bunches of flowers of the most glowing scarlet—the scarlet of the Salvia splendens—diversified with yellow spots. There is also the Melanorrhæa usitata, the Varnish Tree of Ava, allied to the Cashew, but with its fruit surrounded by large six-leaved involucra of a fine rose-colour :—the Aconitum ferox of the Himalaya Mountains, one of the most deadly of vegetable poisons, and used by the Nepalese to poison their weapons :—a great variety of Scitamineous and Orchideous plants of remarkable beauty :—and the curious Apostasia, a genus with the habit of Orchideæ, and in part their structure, but with nearly regular flowers, with two perfect anthers and a pistil constructed like those of ordinary Monocotyledons, and a three-celled ovary.

March 9th. Mr. Grote's motion for the Ballot came on the night before last, but was defeated by a majority of 112, the numbers being 265 to 153. I did not expect it would be carried but I certainly had hoped the minority would have been stronger.

Grote's speech was excellent, and to my mind quite 1837 convincing. The principal speakers on the other side were Spring Rice and Lord Howick.

I spent some time at the British Museum. The collection of organic remains, though not yet completely arranged, is very interesting, particularly the magnificent specimens of Ichthyosauri and Plesiosauri: there is one large head of an Ichthyosaurus very nearly perfect, exhibiting in the most satisfactory manner the curious arrangement of radiating bony plates around the eye, and the other peculiarities described by Buckland in his interesting treatise; an almost entire skeleton of another Ichthyosaurus of a huge size, above 20 feet long, and in a surprisingly perfect state; a very fine specimen of the Plesiosaurus dolichodeirus from Lyme, and another almost equally perfect from Somersetshire; a well-preserved head of a Teleosaurus or Gavial from Whitby, and of a Crocodile from Sheppy, shewing very clearly the difference between the long slender snout (dilated at the end) of the gavial, and the short thick muzzle of the crocodile. There is a fine cast of the pelvis and hinder limbs, and some other separate bones, of the gigantic Megatheorium, so enormously thick and massy that the Elephant's bones appear quite slender and delicate in comparison to them.

Also casts of the jaws and teeth of the strange Deinotherium, described by Buckland, with its tusks bending down from the end of the lower jaw, like hooks by which to suspend itself.

1836   Went in the evening to the meeting of the
Geological Society, where I met Sedgwick, Whewell,
and Murchison, and was introduced to Mr. Darwin
the naturalist, who is lately returned from the
surveying expedition under Captain Fitzroy. I had
hoped to hear an account of some of his discoveries
read this evening, but was disappointed. Sedgwick
made a very amusing speech, in his peculiar style of
rough and ready eloquence mingled with humour,
on the subject of raised beaches, which one would
hardly have supposed to admit of much rhetorical
ornament.

Greenough answered him with more humour than
soundness of argument, and maintained the old
Wernerian notion that the sea had sunk, instead of
the land being raised. Sedgwick, in his reply, justly
characterized this hypothesis as wild and irrational.

March 10*th.* The French Legislature at last shews
some signs of recovering from the state of abject
submission to the Government into which it seemed
to have been thrown by the dread of Jacobin plots.

The long debates in the Chamber of Deputies on
the Disjunction Law as it was called, have ended in
its rejection by a majority of two. This was a
project apparently suggested by the acquittal of the
prisoners of Strasburg, and the substance of it was
that where civil and military men were implicated
together in the same conspiracy or insurrection, the
military should be tried by a court-martial, the others
by the ordinary tribunals. I should suspect that
this was itself intended only as a first step, and that

if it had been carried, the next step would have been 1837 to extend the trial by court-martial to all persons accused of state crimes.

March 11*th*. I met Edward Ellice (the younger) at the Club yesterday evening. He told me that the lessees of Church property in mines, in Durham, express entire satisfaction with Spring Rice's proposed measure touching the Church-rates, but that nevertheless, from the manner in which the Bishops have taken up the opposition to it, there can be little doubt that the House of Lords will throw it out. The King, he says, is again shewing ill-humour towards the Ministry, and was very rude to Archbishop Whately and Lord Plunket at the last levee, while he shewed particular attention to the Bishop of Exeter. All this looks like an approaching change. I am apprehensive that that Church-rate Bill, just and wise as I myself think it, will be the destruction of the present Ministry.

March 14*th*. Went to the Linnean Society, Mr. Don shewed me the specimens of Viola grandiflora in the Linnean Herbarium, which certainly appear quite distinct from our Viola lutea, and differ especially (as Sir J. Smith says) in the much larger flowers, and the much greater proportional length of the spur; the plant altogether is more drawn up, and the flower-stalks longer, than is usual in Viola-lutea, but this may be accidental. The finest and largest sorts of Heartsease in our gardens, he says, belong to Viola Altaica ; some others are varieties of Viola lutea, and others hybrids between those

1837 species.   Hybrids, he thinks, are rarely formed in a
wild state.   He considers the new Heath from
Connemara, called Erica Mackaii, as a variety of
Tetralix; Dr. Hooker seems to think it a hybrid
between that species and Erica ciliaris; the same
has been found near Truro, and also in Spain.   He
told me likewise that the Cow-tree described by
Humboldt has been ascertained to be a species of
Brosimum (of the same natural order as the Fig),
and that its milk consists almost entirely of
Caoutchouc.   The milky juice of plants, he says,
appears always to contain Caoutchouc, and upon
all plants which afford such juice the Silk-
worm will feed; it may be suspected that silk
itself is but a modification of Caoutchouc.   Another
curious thing he told me was, that the upper limit of
vegetation has been found to be higher on the Andes
of Bolivia than on those of Quito, and nearly as high
as on the Himalayas.   Maize is cultivated in Bolivia
at the height of 13,000 feet above the sea, and the
extreme limit of vegetation is between 16,000 and
17,000; the last plant that is found at this elevation
is a species of Culcitium.

March 15*th*.   In the mineral gallery of the British
Museum, I particularly noticed the Sulphate of
Lead, in large brilliant well-defined crystals. transparent and colourless, in cavities of copper-pyrites,
from the Sarys Mine in Anglesea; also in smaller
but equally brilliant crystals, in the cavities of galena
and in still smaller crystals studding the surface of a
ferruginous rock.   There are fine specimens of

Sulphate of Copper, chiefly from Herrengrund in 1837 Hungary, in distinct and rather large crystals, and massive, of a fine blue colour, also in the form of a large stalactite, with a coarse irregular fibrous structure, externally yellowish, and pale blue within. Of Sulphate of Iron there is one very fine specimen, in large, transparent, bright green, rhomboidal crystals, from Bodenmais in Bavaria. The specimens of Carbonate of Copper, both blue and green, are very numerous and of extreme beauty; indeed I am inclined to think this, taking all its varieties, the most beautiful of all minerals. The fine Emeralds from Santa Fé de Bogota are imbedded in white quartz, accompanied by what seems to be black limestone, and sometimes by iron pyrites; there are also some small and pale green emeralds from Egypt, of which the matrix appears to be mica-slate

March 16*th.* The ministerial resolutions on the subject of Church-rates were again brought before the House on Monday last (the 13th), and led to an animated debate of three nights, the division not taking place till half-past two o'clock this morning. Sir Robert Peel led the Opposition attack in a very able speech; Sir W. Follett spoke very well on the same side on the second night, and Lord Stanley on the third. Lord John Russell's reply was admirable. After all, the Ministry obtained but a poor majority on this question,— a majority of only twenty-three in a House of 523 members. I have heard that they expected about forty. Lord Dudley Stuart, and many others who usually support them, went

1837 against them on this question. What astonishes
me is the great number of petitions presented in
favour of Church-rates: I wonder how they can be
got up; I suppose it must be by the exertions of
the country parsons in their parishes, persuading
the ignorant people to sign them without examination, and making them believe that the Ministerial
plan will leave the churches to go to ruin. In
addition to all this, I understand the King has
been completely gained over by the High Church party,
so that I very much fear this measure will be the
ruin of the Ministry. I have heard a story of
Bishop Philpotts. A friend of his, with his family,
was staying with him : in the evening the children
of the two families were to say their prayers together
before going to bed, when, after going through the
usual prayers, the little Philpottsses ended with
praying for Earl Grey. The governess of the
othe children, rather surprised, could not help
asking why they did so;—"Oh," answered one of
them," Papa tells us it is our duty to pray for our
greatest enemies!" The story indeed rests only on
the authority of Sydney Smith, but—"*se non è vero,
è ben trovato.*

March 18*th.* Mr. Roebuck last night brought
forward the question of the seizure of the *Vixen* by a
Russian man-of-war on the Circassian coast, in a
very able, spirited, and manly speech, calling the
attention of the Ministry and the House very
forcibly to the insolent and overbearing conduct of
Russia. Mr. Ewart, the member for Liverpool,

seconded him, shewing the importance of protecting 1837
our trade against such insults. I am becoming
very much reconciled to Mr. Roebuck: I dis-
liked his speech at the beginning of the session,
but on the Irish Corporation Bill, the Poor Law
question, and now on this affair with Russia, he has
spoken with great ability and in a most excellent
spirit.

Lady Holland would have us believe that the
Ministers did not expect a majority of more than
twenty-five on the Church-rate Bill, and were very
well satisfied;—this may be doubted. It is
admitted that most of their followers expected at
least forty.

March 20th. A heavy fall of snow this afternoon.
Mr. Don shewed me a section of an old trunk of
Laburnum, which has a very beautiful appearance;
the ducts and woody fibre are not indiscriminately
mixed, but arranged in alternate concentric layers of
different colours, the zones of ducts being light-
coloured, the woody-fibre nearly black; the medul-
lary rays are numerous, narrow, and very regular,
but between them are a great number of delicate,
wavy, irregularly transverse lines of cellular tissue
giving a beautifully mottled look to the wood.

Mr. Don believes this last peculiarity of structure
to be general in leguminous trees.

March 21st. A great deal more snow has fallen this
morning, and it seems disposed to lie; the roofs are
all white. I am told that the thermometer in the
night was down to 24 degrees.

1837   *March* 28*th*. A fine day, much less cold than it has been lately, the wind having got to the westward of North.   I called on Miss Fox, and had a pleasant chat with her.   Talking of Lord John Russell's success as a speaker, and the way in which his talents have triumphed over natural disadvantages, she told me a droll saying of Sydney Smith's, When Lord John first began to distinguish himself in Parliament, and it was still apprehended that his small frame and feeble voice would hinder his complete success, Sydney Smith said that " he certainly would be the leader of the House, if one could but persuade Sir Watkin Williams Wynn to swallow him ! "—Sir Watkin being a huge, big man.

April 2*nd*. There is a rumour in some of the newspapers that the King, wishing to keep things quiet, and prevent another change of Ministry, has promised to use his influence with the House of Lords in favour of the Irish Corporation Bill, as *that* is the measure on which the Whigs have staked their ministerial existence.   A compromise (if the report be true) is thus to be effected : the Lords are to be induced to pass that great measure, and left to deal as they think fit with the Church-rate Bill, and others less popular or less important.

NOTE.—At the general election that took place in the summer of 1837, at the death of King William the Fourth, and the accession of Queen Victoria to the throne, Charles Bunbury and his father both came forward as candidates for parliament, both on the

Whig side. Sir Henry Bunbury for the western 1838 division of the county of Suffolk; his son for the town of Bury St. Edmunds.

Sir Henry Bunbury and Mr. Wilson the whig candidates were beaten by a large majority. At Bury St. Edmunds a whig and a conservative candidate came in, Lord Charles Fitzroy and Lord Jermyn. The conservative candidate that failed was Mr. Calthorpe. Charles Bunbury the whig candidate lost by a very small majority.

In the autumn of this year 1837, Sir George Napier was appointed governor of the Cape; and he invited Charles Bunbury to accompany him to that colony as his guest; an invitation that was accepted with much pleasure.

Sir Charles Bunbury published through Murray, a Journal of his visit to the Cape Colony, so I only print the letters he wrote to his family during his stay there.

F. J. B.

LETTERS.

Government House,
Cape Town,
January 24th, 1838.

My Dear Father,

We landed last Saturday the 20th, but have been in such an unsettled state ever since, as it were with one foot on sea, and one on land, that I have not till

1838 now been able to sit down to write. We arrived in sixty-six days from Portsmouth, which is considered a fair passage, and though we fell in with a good deal of bad weather in some parts of it, our voyage was on the whole prosperous, and as pleasant as a long voyage can be ; the vessel was very comfortably fitted up, and nothing could exceed the civility and good-nature of the captain. Comparing it with my two former voyages, I should say that this was less prosperous in point of weather than either of them, and yet much pleasanter. My great pleasure was in walking with Sarah and Cissy,* for after the first fortnight they were generally able to come on deck in the afternoon, and many a long and confidential conversation we had together, while pacing up and down the deck. I thought that I knew them both well, and loved them well before, but the being thus almost constantly together for so long a time, has made us still more thoroughly acquainted, and has increased my affection for both of them. Cissy I certainly did not entirely understand or appreciate before this voyage. Poor Sir George himself suffered very much from sea-sickness, but I trust his health is not materially impaired, though he is not so careful of himself as could be wished, and cannot be kept from over-exerting himself ; his spirits, however, are very good since we landed ; John,† I am sorry to say, is not yet in really good health, and requires much

---

* Daughters of Sir George Napier.

† George and John, sons of Sir George Napier

care, but I think he is himself aware of this, and 1838 disposed to be prudent. As for George the younger, with eating, sleeping, and laughing, he has become quite fat. John Craig* I like very much, he has all the Scotch coolness and caution, with the kindest disposition, and an excellent understanding.

We were fortunate in our fellow-passengers, some of whom were really pleasant, and all at least harmless, so that there were none of those quarrels which are said to be common on board of Indiamen.

We were very lucky in getting ashore on Saturday evening, for a 'south-easter' has been blowing ever since, and yesterday with great violence, so that a good deal of the luggage is not yet landed, nor can be till the wind abates. We are pretty well sheltered by trees just here, yet even here we were terribly annoyed yesterday by the dust, but in the town and the open country it was ten times worse—I never saw anything like it. Indeed, the dust seems to be the great nuisance of the Cape at this season; it clogs all one's pores, fills one's eyes, spoils one's clothes, covers books, papers, tables, everything; the very trees in the town look as if they had been washed over with red ochre. Cape Town has in itself no pretensions to beauty, but it is pretty well for a town out of Europe: the principal streets are broad and regular, but unpaved, and many of them shaded by rows of trees, the houses flat-roofed, rather low, and painted of various colours; there are no handsome

---

* Sir George Napier's private secretary and half-brother of his first wife, the mother of his children.

1838 public buildings. The population is a curious
motley mixture of various nations and colours,
English, Dutch, Hottentots, Malays, Negroes, and
Mulattoes, or half castes of every shade—and the
variety is almost as great in costume as in colour.
Among the most singular looking objects are the
long waggons, drawn by twelve or fourteen, or even
more oxen, and driven by a Hottentot, with an
immensely long bamboo whip. You may find in
your library many views which will give you a
general idea of the appearance of Cape Town and
the mountains behind it, for the features of the
scene are so striking and simple that it would be
difficult not to catch the likeness to a certain
degree, but I never saw any view of the much finer
scenery which we first saw on our approach—I mean
the coast running down from Table Bay to the real
Cape of Good Hope. It is a very high and bold
coast, the mountains rising almost precipitously
from the sea, in successive ledges of rock, very
abrupt and ragged, with little verdure, and terminating in rough serrated ridges. They put us all in
mind of the Maritime Alps. The Table Mountain
and Devil's Peak behind Cape Town are fine,
rugged, picturesque fellows, all the upper part of
them appearing (at least as seen from hence) to be
nothing but bare cliffs of stratified rock with the
strata very distinctly marked looking like enormous
slabs piled one upon another. In spite of the vineyards and plantations immediately around the town,
there is a great deficiency of verdure in the whole

scene, and an almost total absence of natural wood. 1838
This, to be sure, is the dryest season of the whole
year, so that the scenery probably looks more
cheerful at other times.

In my next letter I hope I shall be able to tell you
something of Cape politics. In the meantime pray
give my best love to Emily and my brothers — for
Henry I hope is with you before this.

Ever your most affectionate son,
C. J. F. BUNBURY.

P.S. Since I closed this letter, I learn to my
great surprise that I shall see Henry before you do.
George has received a letter from him from the
Mauritius, where he had been delayed a month, he
expects to be here about the beginning of February,
and to stay with us two or three weeks. This is a
most agreeable surprise, as I had confidently
supposed him to be already in England. I think he
will be a little surprised to find me here. I must
also add, that Sarah declares she has not felt so well
for two years as she does now.

Government House,
Cape Town,
February 1st, 1838.

MY DEAR EDWARD,

Though I suppose you will have heard from some
of our party before you receive this, I will not
neglect to write to you by this opportunity, to tell

1838 you my impressions of the place. I was not very
much struck with it at first, though the mountain
scenery is bold and imposing ; it is far from being
so striking or magnificent as that of the Brazilian
coast ; the general naked, parched, barren look of
the country produces an unpleasing effect, and it has
moreover at first sight a much less exotic and un-
European aspect than I expected, owing principally
to the predominance of European trees—oaks and
stone-pines in the gardens and public walks. The
town too with its hot, unpaved, dusty streets, and the
dull sameness of its buildings, makes rather an unfavourable impression. But since I have been
settled in this house, and become more acquainted
with the place I enjoy it mightily. Government
House without the least external beauty is roomy
and comfortable, and its gardens are exceedingly
pleasant, full of fine trees and curious plants of
various countries, and affording a shade and verdure
which are delicious at this season. The trees too
protect us in a great measure from the south-
easterly gales and dust which are at this time of
year the great nuisances of the Cape, and are
unmatched in their respective ways by anything I
have felt elsewhere. The ' south-easters,' however,
furious and annoying as they are, are very healthy,
and without them the place would be intolerably
hot in the summer, backed as it is by a wall of
naked mountains facing the noonday sun. As it is
the thermometer has more than once been above 90
degrees in the shade since our arrival ; but I do not

find the heat oppressive within doors, and the girls 1838
delight in it.   When the dust is not flying the
weather is certainly delightful, especially in the
mornings and evenings, and even Cissy allows that
the air is as clear and the sky as bright as in her
dear Italy.   We have from the garden here a
beautiful view of the Devil's Mountain and Table
Mountain, fine, bold, rugged fellows, particularly the
latter which looks like the shattered wall of some
gigantic, ruined fortress with its magnificent stratified cliffs, rent by deep ravines and fissures.   The
lights and shadows on the rough sides of these
mountains, in the morning and evening are remarkably beautiful.   The ' south-easters ' are almost
always accompanied by a peculiar cloud which they
call the Tablecloth, a long, dense, white cloud,
looking like a wreath of snow which lies along the
top of the Table Mountain, and remains stationary
there sometimes for days together, while the sky
everywhere else is perfectly clear.   The general
appearance of the country about Cape Town is very
bare, and parched, and dusty ,though relieved by occasional gardens, vineyards and plantations of oak, pine,
and Silver-tree (this last seems to be the only indigenous
tree of the place, and is also largely cultivated both
for ornament and firewood).   The vines are cut down
even lower than in Italy and have no picturesqueness
or beauty beyond their brilliant verdure.

So much for the scenery of the Cape.   I suppose
you will have received from Sarah or Cissy a
sufficient bulletin of health, so I shall say nothing

1838 on that subject, except that I never saw those two dear girls look better than they do now. My chief enjoyment during the voyage was walking and talking with them, and I find my affection for both of them increase more and more as I know them more intimately. I read on the voyage a great part of Spenser's " Fairy Queen ", which I had often felt ashamed to be so ignorant of. I do not mean that I read it regularly through—*that* would far exceed my patience, and I should think most people's—for it is the most involved, perplexing, complicated allegory, as well as the longest, I ever met with, and the characters are as uninteresting as the story. But the fertility and brilliancy of his imagination, the richness and power of his descriptions, and the variety and picturesque effect of his personification, are really wonderful. What I most dislike in him is the too frequent introduction of disgusting images into his descriptions. After all if he is to be ranked (as he usually is) next to Milton, I think it is at a very long interval.

We are expecting Henry here very soon, on his way from the Mauritius (whence he wrote to George) and I hope he will stay three weeks or a month with us. I suppose you are by this time settled in chambers, and hard at work on the law ; I wish you all possible success, and I hope you will like London as well as I did when I spent a winter there. God bless you, my dear brother.

Believe me ever,

Yours most affectionately,

C. J. F. BUNBURY.

Government House,   1838
February 2nd, 1838

My Dear Father,

I wrote to you on the 24th of last month, giving
an account of our voyage, and of my first impressions of the Cape: this is number two. Henry
arrived yesterday, sooner than we had expected ; he
is looking very well and as you may suppose, not
grown any shorter than he was ; he is lean but hardy-
looking, excessively brown, with formidable black
eyebrows and whiskers, and hair still more grey than
mine. He is full of interesting stories and information about Australia, and what surprises me most is,
that he seems to have grown after all quite fond of
Swan River. George*, though not yet perfectly well,
is getting much better ; and is very cheerful and seems
to enjoy the business of his office. The two girls are
uncommonly well,—Sarah over head and ears in
household cares and business, to which she seems to
take very kindly, and makes an excellent house-
keeper. John, the only one of the party whose
state of health is calculated to give real uneasiness,
is very cautious and prudent. I have seen a good
deal of Colonel† and Lady Catherine‡ Bell, and like
them very much: Lady Catherine, with a certain
austerity of look and manner at first which is not
encouraging, is a very kind, as well as a very sensible
and agreeable woman. Colonel Bell seems to be

* Sir George Napier, Governor of the Cape.
† Afterwards Sir John Bell.
‡ Daughter of the Earl of Malmesbury.

1838 universally respected in the colony, and I should think very deservedly; his mild, sedate manner and benevolent look make a very favourable impression upon one, and I have been much interested by his conversation; he is not only a sensible man, but seems to be most thoroughly acquainted with everything relating to this colony. I am not sure that he is quite unprejudiced with respect to the question between the colonists and natives; one day that I dined with him he talked with great vehemence on that subject, taking up very warmly the cause of the Dutch colonists, who, he declared, had been most unjustly abused and misrepresented. After all it is very possible that the statements on the other side may be equally prejudiced. My plan is, as far as I can manage it, to express no opinion of my own on such subjects, and to say only so much as is necessary to draw out those of others, and thus learn all I can.

Colonel Smith* is a fine, frank, determined-looking soldier, and I apprehend from what I learn here that he has been very unjustly censured for harshness and severity towards the Caffers, though it is true that he in great measure brought himself into the scrape, by the wild rodomontading style of his dispatches. The strongest testimony in his favour, I think, is that of Sir John Herschel, who is not likely to be tainted with colonial prejudices, and who speaks of him as a most humane, and

---

* Afterwards General Sir Harry Smith

excellent man. Besides, it is admitted that he 1838 gained an extraordinary influence and ascendancy over the Caffers, while he was in command of the province of Adelaide. I have called upon Sir Benjamin D'Urban, the late governor, who is now living in a cottage at Wynberg; he enquired much after you, and seemed much pleased by your remembrance of him. He is in great distress at present, poor man, having lost his favourite grandson by a shocking accident a few months ago. What I hear of his government here inclines me to think that he was a very well-meaning man, but without much political ability or decision, easily warped or misled by designing or prejudiced men; he is much censured for having delayed so long at Cape Town, instead of hastening to the frontier, as most people seem to think that if he had gone there earlier, and looked into the state of affairs with his own eyes, the Caffer war might have been prevented. I have told you freely my present impression of those people,—not as deliberate opinions, but impressions which may very possibly be mistaken, and are open to correction. The great subject of discussion here, at present, is a Sunday Bill which Sir Benjamin was prevailed upon (by the missionaries and the rigid Presbyterian party) to pass, and which is nearly as bad as Sir Andrew Agnew's, forbidding the sale, even of meat and bread on Sunday : it has excited a great deal of discontent, and a strong and probably successful attack will be made upon it in the Legislative Council. On the other hand, the amend-

R

1838 ment goes a little too far, I think in permitting *voluntary field labour* on the Sunday which would be very apt to become involuntary labour. George has promised to take me to the meeting of the Legislative Council to-morrow, to hear the question discussed, but as the ship by which this letter goes is to sail to-morrow, I shall not be able to send you any account of the debate this time. It will be a great pleasure whenever a fresh supply of newspapers arrives from England, to bring us the account of the first proceedings of Parliament.

This is the worst season of the year for botany, as a great proportion of the plants are withered and burnt up by the continued heat and drought, yet still there are many interesting plants in flower, particularly Heaths and other shrubs of a similar appearance. Everlastings and Papilionaceous-flowered shrubs. I have been out several times on the heaths between Cape Town and the Table mountain, and the variety of species to be met with in a small space even at this season, is very striking. There are a great many of the Protea tribe, but only one in blossom at present, and that not one of the handsomest. The Silver-tree, which belongs to that tribe, seems to be the only indigenous tree of this corner of the colony, it is remarkable for the brilliant white satiny down which clothes its leaves, making them look really like silver, and producing a beautiful effect when they are shaken by the wind; it is much cultivated for ornament, but it is of no known use except as fire-wood. The most abundant trees in and around Cape Town, however, are the 1838 common oak and the stone pine, both introduced from Europe, and thriving exceedingly. The oak however, as Colonel Bell tells me, both grows and decays much faster than in England. I pass my time here as pleasantly as possible ; George has good naturedly assigned to me a very snug little quiet room looking into the garden, where I can read and write, dissect and dry plants, draw and muse, as much as I like. The house is spacious and comfortable, and the gardens delightful, even in this parching weather they have shade and verdure, and the trees shelter us in a great degree from the south east winds and the dust. The weather is very hot, but agrees exceedingly well with me.

Henry has the most gratifying letters of thanks and approbation from the Governor and the Major Commandant of Swan River, and an address of thanks from the principal inhabitants of the colony.

I hope that your health and Emily's have not suffered from the winter and that you are going on prosperously with Sir Thomas Hanmer.* I write to Edward by this post. Pray give my kind love to Emily, Hanmer, Mrs. Gwyn, Miss Fox, etc., and

Believe me,
Your very affectionate son,
C. J. F. BUNBURY.

* The Life of Sir Thomas Hanmer, the Editor of Shakespeare and Speaker of the House of Commons in the reign of Queen Anne.

1838   Government House,
                                           Cape Town,
                                      February 15th, 1838.

MY DEAR FATHER.

Since I last wrote to you (which was on the second
of this month, immediately after Henry's arrival) I
have seen newspapers from England, by which I was
very glad to learn that the Ministers had begun
their campaign so prosperously, knocking down the
Radicals with one hand and the Tories with the
other,—that the Duke of Wellington was disposed to
be conciliatory, and that the Ministry had proposed
an enquiry into the Pension List. This last measure
will be very popular, if I may judge by the fuss that
my own constituents (that are to be) made about the
question. I confess I think Lord John Russell wrong
in his opposition to the ballot, &c., but Mr. Wakely
was ten times more wrong in thrusting those
questions forward so unreasonably and rashly.—
Here, there has been much talk occasioned by
Lord Glenelg's new regulations touching the Legislative Council,— one, that the unofficial members
shall hold their seats during Her Majesty's pleasure
instead of for life ; the other that no member shall
bring forward any measure or question, public or
private, himself, but must lay his project before the
Governor, in whom every measure must ostensibly
originate, and who may, if he chooses, quash it
without discussion. The first of these is surely an
alternative for the better, even though it makes the
Council wholly dependent on the Government; the

second is startling, and seems despotic enough, but 1838 its effect is to concentrate all the responsibility of measures in the Governor, and as the members of the Council are *not representatives*, the mischief may not be so great as it seems. The said unofficial members, however, and those which set up for being par excellence the *Colonial* newspapers are in great wrath. There are three newspapers published here, besides the "Graham's Town Journal," the oldest and much the ablest of them is the "Commercial Advertiser," conducted by Mr. Fairbairn, who made himself conspicuous first by his opposition to the tyranny of Lord Charles Somerset, and since by his zeal and perseverance in the cause of the illused natives. His paper is strongly liberal, without being violent, and very well written; and he himself is highly spoken of by impartial men. To the "Colonial" party he is nearly what O'Connell is to the Irish Orangemen. By the bye, perhaps you are not acquainted with the nature of the Legislative Council. It was first established when Sir Benjamin D'Urban came out as Governor; it consists of 5 ex-officio members besides the Governor, and 5 (or more, for this is not limited) unofficial members, nominated by the government, and (now) removable by it at will. So you see that it is in no degree a representative body, and serves only to advise the Governor (without being able to control him), and to bring the local knowledge and experience which he is not so likely to possess. Even if the Council reject a measure, the Governor may, if he think fit, enforce it

1838 *provisionally*, subject to the approval or disapproval of the ministry at home.

February 16*th*. I have this morning received your letter of the 28th November, by the Semiramis steamer, just arrived. I am a good deal surprised at Bessie's* marriage, but at the same time glad, for I think it will please my uncle and aunt very much, and I see no reason against the marriage being a happy one; all that I do know of Lord Arran (which is but little) is in his favour. I have made much acquaintance with Mr. Harvey, the Treasurer here, a friend of Mr. Spring Rice's. an excellent and zealous botanist and a very agreeable man. He is generally considered as uncommonly shy and difficult to be acquainted with, but we get on famously together, from having a common object of interest and pursuit. We had arranged a plan for going up the Table Mountain together yesterday morning, and the preceding day promised everything favourable, but behold, a wicked south-easterly wind got up in the night, and when *we* got up in the morning, it was blowing hard, and the mountain had got his " Table-cloth " on, so our project was no go. It is difficult at this season to catch a favourable opportunity for going up the mountain, for it is dangerous and almost impracticable when the south-easter blows. Henry is still with us, and I think in no hurry to go. He is a very fine fellow, I think, calculated to do great things; he has an uncommon talent for

---

* Elizabeth Napier, daughter of Sir William Napier, married to the Earl of Arran.

observation and exploration, nothing escapes him, 1838 and nothing seems able to fatigue or daunt him. His manners are still as good as if he had lived all this time in the very best society, instead of among kangaroos and convicts.—I think all our party are pretty well except poor John, who is suffering much, but bears it with a resolute and patient spirit, his complaint I understand is sciatica. I have met Sir John Herschel two or three times. Lady Herschel is a remarkably pleasing woman.

February 21*st*. From what I have since read in the newspapers, I believe I was too hasty in the rejoicings with which I began this letter, as to the situation of the Ministers. They seem to have thoroughly disgusted and estranged the ultra-Radicals, so that their tenure of office is at the mercy of the Tories, who however appear to be in no great hurry to take advantage of their weakness. But it is difficult to judge of the true situation and conduct of political parties at such a distance.—I have been staying a couple of days at Colonel Bills, 5 miles out of town, where I collected a quantity of plants and picked up a good deal of curious information—but I have not room for it here. The Melville is not yet come in. George has given several large dinners, which have gone off very well, though of course they were not very amusing, as great formal ceremonious dinners never are, however they have given great satisfaction I understand, especially to the Dutch (who thought themselves neglected by Sir Benjamin and Lady D'Urban) and George and his girls are on the

1838 high road to popularity, without *courting* it. The prudence and good sense and discretion which Sarah shows in her new and embarrassing situation of dignity, are beyond what I could have conceived· I beg to observe that this is the third double letter I have written to you in one month from our landing here, so that of the quantity at least of my correspondence you have no reason to complain. My best love to Emily and my brothers.

Ever your very affectionate son,
C. J. F. BUNBURY.

Government House,
Cape Town,
March 8th, 1838.

MY DEAR EDWARD,

I take the opportunity of the Windsor sailing for England to write you a few lines, and particularly to give you some account of my ascent of Table Mountain. I had 3 or 4 times been disappointed by south-easters, which come on very suddenly and make the ascent of the mountain both difficult and dangerous, but at length on the last day of February I succeeded in getting to the top in company with Mr. Harvey, the Treasurer of the Colony, an excellent botanist and very agreeable man, with whom I have lately become very intimate. We started at 4 in the morning, and were two hours and a half going up, most of the time good hard climbing; there is no part of the ascent so bad as the Cheminée on the

Brevent, no place when you are obliged to trust more 1838 to hands than to feet ; but for mere foot-climbing it is about as steep as I ever saw, and mighty rough into the bargain, the principal part of the ascent being over loose sharp stones and fragments of rock. Being much out of the habit of climbing, I was quite stiff for a couple of days after it. We stayed 3 hours on the top, where we collected some very beautiful plants (and where by the bye it was very cold)—and were just beginning to descend when a strong south-easter came on, the mountain was presently covered with clouds, and we got wet to the skin in going down. The suddenness with which these clouds come over the mountain, especially at this season, is what makes it very necessary to be cautious, lest one should be caught in the mist on the summit, while far from the pass, for there is only one practicable way to ascend or descend by. Some people have been killed by losing their way thus, and falling down the cliffs, and many have been obliged to spend a night on the top without shelter. However Harvey and I got down without any accident, having been about 9 hours on foot. The height of the mountain is nothing at all remarkable, being hardly greater than that of Snowdon (3582 feet) but it has a more imposing appearance than Snowdon from its excessive steepness, all the upper part, for more than half way down, being sheer cliffs of hard quartz-sandstone, in very thick horizontal strata. The way by which one goes up is a deep gully or cleft in these rocks, tolerably wide at the bottom, and

1838 narrowing to a few feet at the top, where it opens out on the table land. It is a mountain very rich in plants, but I think it would hardly repay anyone but a botanist for the trouble of the ascent, for the view is insignificant. Henry is going home in the Windsor, which sails the day after tomorrow ; he has been with us about 5 weeks, having arrived the day after I last wrote to you. He is a very clever and fine fellow, improved both in sense and manner since I saw him in England (which was 5 years and a half ago, before my first visit to Freshford) and is full of interesting information about Australia, especially touching the natives, with whom he seems to have made himself better acquainted than most Europeans. Sir George during the last month has given a number of great formal dinners. Sarah shows a wonderful talent for housekeeping, and George the younger an equally great talent for doing nothing, which seems to be the chief business of an Aide de Camp. There a few people here whom I like very much, besides our own party.—Mr. Harvey, whom I mentioned before, Colonel Smith, Colonel and Lady Catherine Bell, and their nephew Charles Bell, who travelled in the interior with Dr. Smith*, a very clever, acute, entertaining man. I do not think however I should like this place as much as Rio, if I were not here under such peculiarly favourable circumstances, with such a pleasant party and so forth : there is not much natural beauty, and above all there is not that peculiar charm of

---

* Doctor Andrew Smith.

strangeness and exoticality (to coin a word for the 1838 occasion) which made such an impression on me at Rio, nor as yet do I perceive an extraordinary merit in the climate. During most part of February the weather was very disagreeable, furious south-easters alternating with days of oppressive heat—but it is now improving. I believe we shall set out for the frontier about the end of this month, that is to say, His Excellency, Major Chartres, young George and myself. I am delighted to hear from Sarah that you have taken so zealously to the law, and find it so intersting, and I daresay as you have begun so well, you will go on triumphantly, and find all its difficulties give way before you. I wish we could send you some of our spare sunshine, which you would probably find very agreeable in London. This Canadian War is a sad thing. I daresay the Tory newspapers make the worst of it in order to embarrass the Ministry, but still I do not like the looks of it at all.

Ever your most affectionate brother,

C. J. F. BUNBURY.

———

Government House,
March 16th, 1838.

MY DEAR FATHER,

I thank you much for your letter of December 21st, which I received yesterday, and for send-ing me newspapers, some of which I have read with a great deal of interest. Lord Mulgrave's

1838 vindication of his government is admirable, and most convincing. We are to set out for the Eastern frontier on Thursday next—we being his Excellency, Major Chartres, Dr. Clark of the 72nd, George and myself; I suppose we shall be about four months absent, in which time we shall probably see all that is worth seeing in the Colony, and I hope to make large collections.

We have four waggons, and I do not know how many horses, all these being supplied by the Government.

There have lately been some alarms on the frontier, arising chiefly out of a mutiny in the Hottentot Cape corps, which it was suspected might have been instigated by some of the Caffer chiefs, and Colonel Somerset's injudicious measures seem to have aggravated the alarm, for which the Lieutenant Governor Stockenstrom seems to think there is no real ground; it has, however, increased the Governor's anxiety to proceed at once to the quarter where these apprehensions exist. He is the more anxious, as much blame is thrown on Sir Benjamin D'Urban for his dilatoriness under somewhat similar circumstances, for many people think that the last Caffer war might have been prevented, if *he* had visited the frontier in time. The mania of emigration still continues and appears to be spreading among the Boors: the causes of it seem to be manifold,— dislike to the Slave Emancipation act, and a fear of being deserted by all their apprentices as soon as the time for their liberation arrives,—dread of fresh attacks from the Caffers, a notion that the British 1838 Government is under the control of the missionaries (whom they mortally hate) and that they can expect not justice from it,—exaggerated ideas of the fertility of the country along the east coast—probably also a wish to escape from the restraints of law, and the burden of taxes. I understand that the great merchants in Cape Town have made on speculation very large purchases of the lands which these Boors abandon, and that one of them at least has a project of bringing out a number of German colonists to settle there. I heard a report yesterday that Dirgaan, the great chief of the Natal country, in which the emigrant Boors have settled, after a show of good-will and encouragement to them at first, had attacked them by surprise and massacred 250 of them. This however is a mere rumour, which wants confirmation, though it does not seem unlikely.

Since I last wrote to you I have been up to the top of Table Mountain, with my friend Mr. Harvey, and being out of practice in climbing, I was quite stiff for a couple of days after. The mountain is not higher than Snowdon but exceedingly steep, and the way up is over loose stones and sharp pieces of rock. We started at 4 o'clock in the morning, were two hours and a half going up, and stayed three hours on the top, where we collected a great number of interesting plants, in particular an Orchis* of extraordinary beauty, which is quite peculiar to that spot. We got wet to the skin in coming down again. Mr. Harvey

* Disa.

1838 is one of the few people here (besides our own party) whom I like very much. I have the highest regard for Colonel and Lady Catherine Bell, and Colonel Smith and Charles Bell are extremely intelligent and agreeable, but with these exceptions I have as yet seen no reason to be charmed with the Society of the Cape.

Party spirit seems to be almost as bitter as in Ireland.

February 15th—21st. Henry sailed last Sunday in the Windsor, in company with Sir John and Lady Herschel, and I suppose he will be arrived in England before you receive this. The Melville arrived a few days before he left us, and Major Chartres is with us now; I like him much. I am very sorry that the Canadian quarrel has broken into open war; even if we succeed in putting down this insurrection, as I hope we shall, I apprehend that we shall never be able to hold that country (Lower Canada at least) except as a merely military dependency, which would surely be of little advantage. It will then, I think, be a question very open to discussion whether it is not better to throw off than to retain a colony ripe and ready for independence. The debate on the Pension List is very interesting, and some of the speeches on both sides brilliant and powerful. Pray give my best love to Emily, and to Mrs. Gywn and Miss Fox if this finds you in London. I wrote to Edward on the eighth.

From your affectionate son,

C. J. F. Bunbury.

Graham's Town,
April 16th, 1838.   1838

My Dear Father,

We arrived here the day before yesterday, after a
fatiguing and in part disagreeable journey of above
600 miles from Cape Town.   We were in all 24 days
on the road, of which however only 17 were actual
travelling. we were detained 2 days by rain and
swollen rivers, stayed 2 days in Port Elizabeth, and
one in Uitenhage, and were besides stationary 2
Sundays, as it is not the custom in the Colony to
travel on that day   These days of rest were the
only ones I enjoyed at all, the others, generally
speaking, we started very early in the morning,
travelled some hours before breakfast, and after
jogging all day over horrible roads and ugly country,
arrived at our night's quarters too tired to examine or
dry any plants, even if I had leisure to collect any on
the road.   The country in general is barren, dreary,
and monotonous,—arid plains without a tree or a
spot of cultivated land, and mountains as naked as
the plains, though often picturesque in shape.
There are however some pleasing spots : Zwellendam
and Uitenhage are pretty villages, and prettily
situated, but still inferior to a great many places in
England.   On this side of the Camtoos river there is
a great extent of *bush*, a singular kind of country,
and not without beauty ; it is not exactly forest, for
there are no large trees, and indeed few that deserve
the name of trees at all ; it is a tangle of thick
bushy shrubs and succulent plants, seldom rising as

1838 much as ten feet high, most of them thorny, and generally loaded in a strange fantastic way with long tassels of a grey thready lichen. These thickets, which are very difficult for man to penetrate, still afford a shelter to lions and elephants, which have been extirpated from most other parts of the colony. Aloes, Euphorbias, and other succulent plants abound particularly in these eastern districts, which have a vegetation in general very different from that of the Cape. I have had some good botanizing days, especially at Zwellendam, Jagersbosch, and Port Elizabeth, but on the whole my collections have fallen far short of my expectations, partly owing to the drought, but much more to our hurried travelling. You may be surprised perhaps that I should complain of the drought, after having said that we were delayed 2 days by rain ; but the fact is that it rained only one of those days, and we were assured that there had not been such another day of settled rain for the last 2 or 3 years. The second day we were stopped by the swollen state of a river which we had to cross. Everywhere we hear bitter complaints of the droughts. In some districts we have passed through, the people have not been able to plough or to sow grain for two years past, for want of rain, and the cattle and horses are generally in a very poor condition from the same cause.

With respect to the Dutch farmers of the colony, I should be shy of giving a decided opinion about them, having passed in a hurried manner through their country, without understanding their language, and seeing them moreover, probably, in their best 1838 colours, as I was in the company of the Governor. I may mention however my impressions of them, corroborated by what I have learnt from the Lieutenant-Governor Stockenstrom, and others well acquainted with the country. They seem to be hospitable, much disposed to be civil and kind to strangers, and ready to oblige (if they can do so without much trouble to themselves),—brave, sincere, and simple in their habits ; but ignorant, indolent, extremely apathetic, and attached to a routine from which scarcely anything can induce them to deviate.

In almost every farm-house one sees a large family of grown-up sons and daughters, who lounge about in idleness, leaving all the work to their Hottentots and apprentices, learning nothing and doing nothing, till on the death of the father all his property is equally divided between them. Whatever may have been the case formerly, they do not now appear disposed to be unjust or cruel towards the native tribes. The English settlers of Albany are far more inveterate in their hostility against these.

We have seen in the course of our journey, many instances of the prevailing mania for emigration and have heard much on the subject. In my last letter (of March the 16th) I gave you such information as I had been able to collect touching the causes of this movement, but I have since been led to think that the reasons I then gave you were *Cape Town* reasons, and had had but little, if any, influence on the great mass of the emigrants. In the country

s

1838 and among the neighbours and acquaintances of the emigrants, one hears much more simple and natural explanations. It is very likely that disaffection to the British government, and dislike to the system of policy lately adopted, may have had some weight with the *leaders* of the emigration, but the great majority seem to have been influenced chiefly by some vague expectation of bettering their condition, escaping from the distress occasioned by the long drought, and finding a more fertile and agreeable country. With a great many, too, the ruling motive was merely the spirit of imitation, the instinct of following the leader. They say—" Our neighbours So-and-So are gone or going, we must do as others do." After all, emigration in seasons of distress is no new thing, though it never took place before on so large a scale, or with so much method. ·In the old times of the colony, the frontier farmers used to emigrate and take possession of new lands whenever they found themselves distressed by drought or want of room ; but then the limits of the colony were extended in proportion as they pushed on ; now, the limits are fixed, and they therefore remove out of our jurisdiction. The worst of it is, that these emigrants have carried an enormous quantity of cattle and no small sum of money out of the colony. They have also decoyed or forced a great many of their apprentices over the frontier, and when once out of the reach of British law, have reduced them again to slavery. I mentioned in my last letter a report that many of the emigrant boors had been massacred by Dingaan, the Zoola chief: this turns 1838 out to be true. Dingaan (influenced, as he says himself, by the fear that the Boors would overpower him, and take possession of his country—not an unreasonable fear it must be owned) decoyed Retief, the original leader of the emigration, and 60 of his men, into a snare, and slaughtered them all; then he attacked the camp by surprise, and butchered, it is said, 250 more, including women and children. The main body of the Boors, however, being quartered separately, escaped this attack, and afterwards set out on a *commando* against him, of which the result is not yet known. It is likely to be a dreadfully bloody war, however it may end. But this is not all. A colony of English have established themselves at Port Natal, without any sanction or authority from the Home Government;—these, on hearing of the massacre, set out on a command, ostensibly to punish Dingaan, but in their way they attacked another tribe and carried off several hundreds of women and children as slaves, besides a multitude of cattle. These are a pretty sample of our countrymen in the southern hemisphere. There is in the *Graham's Town Journal* a curious account of the massacre of Retief, taken from the journal of a missionary who was at the time resident at Dingaan's Kraal; that chief seems to be an uncommonly shrewd and crafty savage, and to have reasoned very acutely, from the conduct of the European settlers towards other native tribes, as to his own probable fate. There was not long ago a

1838 serious mutiny in a company of the Cape corps
stationed on this frontier, they fired into a hut where
several of their officers were sitting, and killed one.
Fourteen of the men are now under sentence of
death, and I believe four will be executed. If the
confessions of the mutineers may be believed, their
plot had been concerted with Umkai and some other
Caffer chiefs, who were to attack the colony at the
same time, but something or other caused it to
explode prematurely. It has thrown the Graham's
Town people into a dreadful fright, notwithstanding
the presence of the 72nd regiment. These worthy
settlers are excessive poltroons with regard to the
Caffers.

The Lieutenant-Governor Stockenstrom has been
with us for the last fortnight. He is a remarkable
man : there is a vehemence, and earnestness, and
something declamatory in his manner, which does
not give one the idea of impartiality, and he is
perhaps too imperative and irritable for the peculiar
situation in which he is placed ; but I am convinced
that he is a thoroughly upright and sincere man, and
vehement only in what he conscientiously believes
to be a good cause. He is undoubtedly an able
man, and thoroughly well acquainted with the people
of this colony, and everything relating to it. Nor is
he one of those who would do justice to the natives
only, and sacrifice the just rights and interests of the
colonists. Stockenstrom, as far as I can perceive, is
anxious to do full and equal justice to all, and to
promote the real interests of the colony by maintain-

ing peace and good understanding with the native 1838 tribes. He is most rancorously hated by the Albany people, and the so-called " colonial " party, because having in his youth acted with them, and shared their feelings of enmity towards the Caffers, he has seen his error, and become the champion of a better system. I do not know if you see any Cape papers. You will of course not put your faith in the *Graham's Town Journal*, which is just about as true and fair in its representations as an Irish Orange newspaper ; but I hope you will not give implicit credit to Dr. Philip and his missionary set, whose statements are a good deal distorted and coloured by party and sectarian zeal. I think I mentioned before that the *Commercial Adviser* is by far the best of the Cape papers.

As I began my letter with complaining of the fatigue of the journey, it may be necessary to add that I am nevertheless in perfect good health. I am sorry to say the Governor is not quite so well : he is feverish, and seems to suffer a good deal from the fatigue and heat ; but his good nature and cheerfulness are unvarying. I have kept a journal, though I do not know that observations made on such a hurried journey are much to be depended on. By-the-bye I have omitted to tell you our route ; if you look at Arrowsmith's new map of the Cape, you may easily trace it, by Hottentot Holland, Sir Lowry's Pass, Houw Hock Pass, Caledon, Zwellendam, across the Gauritz river near Heuringklip, by Attaquas Kloof, down the Long Kloof to Jagersbosch on the

1838 Kromme river, thence to the mouth of the Camtoos, Port Elizabeth, Uitenhage, Addo Drift, Rottenburg's Drift, Graham's Town. Next week we go to meet the Caffer chiefs beyond the Great Fish river.

My best love to Emily and my brothers. I hope Edward has remembered to tell the secretary of our Club that I am abroad, otherwise I shall be considered as a defaulter.

Ever your most affectionate son,
C. J. F. BUNBURY.

––––

Graham's Town,

May 16th, 1838.

MY DEAR FATHER.

We returned about a week ago from a ten days excursion into Cafferland, which proved by far the most interesting part of our whole journey. It would be almost useless to give you an itinerary of this tour, as few of the stations are marked in Arrowsmith's map, or in any published one that I have seen. I was present at two conferences between the Governor and the Caffers,—one at the head of the Beeka river, near Fort Peddie, where the chiefs of the Congo tribe—Pato, Kama, and Cobus, and those of T'Slambie's tribe—Umhala, Gazela, Umkai, and Noneebe, met him with above 300 of his followers,— the other at Block Drift, on the Chumie, where Macomo, Tyali, and the other chiefs of Gaika's tribe

attended. I have never seen anything more interest- 1838
ing than these meetings : the second was the more
striking and picturesque of the two, for it was held
in the open air, in a green meadow studded with large
acacias, where the Caffers, nearly 300 in number,
seated themselves in a semi-circle around the little
group of English who attended the Governor and
Lieutenant-Governor ; and the wild appearance of
these fine barbarians, their dark and athletic figures
either quite naked or partially covered by their
Karosses (cloaks of wild beasts' skins) their animated
intelligent countenances and looks of earnest attention, their strange ornaments and weapons, and the
showy uniforms of the English officers contrasting
with the dusky natives, together with the scattered
groups here and there under the trees, the tents, the
horses, grazing, and the rich verdure of the spot,
made altogether the most striking scene I ever
beheld. The other conference took place in a
missionary church, and was therefore less picturesque ; besides, the discussion was less animated.
Both Conferences terminated very amicably and all
the chiefs concurred in professing the most pacific
intentions, and the greatest aversion to war,—how
far sincerely, God knows ; but I do not think it can
be their interest to plunge into a fresh war without
provocation. Even through the medium of an interpretation, we were all struck with the tact and
acuteness and dexterity which the Caffers displayed
in these discussions, and which would have done no
discredit to the most accomplished debater in the

1838 House of Commons. I never saw a public meeting
in England so orderly as these two: of all the
numbers present, not one interrupted the person who
happened to be speaking, but all listened in silence,
and with the utmost attention. The two famous
chiefs Macomo and Tyali, who were the prime
leaders of the late war, dined in company with us at
the house of Mr. Stretch (the resident agent of our
Government in Cafferland), and behaved quite like
gentlemen. They conversed very readily by means
of an interpreter, and showed extraordinary quickness
and acuteness in their remarks and answers. I do
not think that the appearance of the chiefs is improved by the European dresses which they wear on
these occasions as a mark of honour: they would
look better in their karosses ; but perhaps the value
and the idea of dignity attached to these dresses
may be regarded as a sign of approximation to civilized tastes and feelings. The Caffers are so far from
"rapidly disappearing from the face of the earth,"
as Sir W. Molesworth seems to assume in his
speech on the colonies, that they are now far
more powerful and formidable than when the
English first came in contact with them, and are
likely to go on gaining ground. They have now a great
quantity of horses, and know how to use them : it is
estimated that Macomo can muster 360 mounted
men, and the T'Slambie tribe as many more. Many
of them are provided with fire-arms, and the frontier
colonists, notwithstanding their dread and hatred of
the Caffers, are yet induced by the "lucre of gain"

to carry on an active contraband trade with them in 1838 muskets and ammunition. It is true that few of them are as yet skilful in the use of these weapons, and that, at present, the musket is probably less formidable in their hands than the assegai ; but there is no reason why they should not in time become as good shots as the N. American Indians and they will then be truly dangerous enemies in the bush. It is clear to me that there is no choice, but either to keep up a peaceful and friendly intercourse with them, which I firmly believe to be practicable,—or else (if any Government could be wicked enough to adopt such an alternative) to crush them at once and without delay ; for in less than another century they will probably be too strong for the colony. The chief obstacle to the maintenance of peace with them is their inveterate habit of thieving, which they look upon as no crime, but rather a creditable exploit. They are as determined and dexterous cattle-stealers as the Highland " Caterans " or the Borderers whom Walter Scott celebrates ; and the nature of the country along the frontier line is peculiarly favour- able to their enterprises. The chiefs profess a willingness to check these depredations, but they do, and perhaps can do, little or nothing, and the colonists along the border are exposed to severe losses, and kept in a constant state of irritation and anxiety. No doubt their complaints are in many respects unreasonable or exaggerated, and I believe also that their losses are often occasioned by negligence and want of proper precaution on their

1838 own part; but still it is certain that they do suffer severely from the depredations of the Caffers. It is not easy to devise a remedy for this evil. The Governor is about to adopt a new and more effectual system of patrols, and has also made an arrangement with the Gaika chiefs by which it is supposed the punishment of the marauders will be made more efficacious. But I do not expect that any measures will be sufficient to put a stop at once to a practice so rooted in the habits of the race. For many years, most probably, the border colonists will continue to suffer from depredations. How can they expect otherwise, living on such a wild frontier, in the immediate neighbourhood of a bold, crafty, uncivilized race, of inveterate predatory habits? Anything is better than a return to the old system of commandos, which, without at all preventing thefts, kept up a continually increasing feeling of soreness and ill-will and enmity on both sides, naturally terminating in a war. I do not say however that the colonists (especially the Dutch) have no real grievances, unreasonable as they are in their clamours against the present system of policy. They have received no compensation for their losses by the late war, and, what is much harder, no compensation whatever for the cattle and horses which they were compelled to give up for the use of the army. Another thing they complain much of, and not without reason, is the annoyance and loss they are subjected to by the numbers of vagrants, Hottentots, Fingoes, and others—who roam about the country in all directions without any means of 1838 livelihood, and beg or steal as occasion serves. But it is said that anything like a vagrancy law would be impracticable in this colony.

The Governor has too much business on his hands to be able to write to you. He works tremendously hard, but keeps his health, I am happy to say, and I see continually more and more reason to admire his manly fearlessness and straightforwardness, and singleness of purpose, the soundness of his judgment, and the excellence of his heart.

May 17*th*. I have to-day received your letter of February 8th, and thank you very much for it. I am particularly glad to know your opinion of the Canada business, and to find that it agrees with that which I had formed from other sources. I am sorry that Grote, whom I rather admired, has mixed himself up with the advocates of rebellion. I have seen an English newspaper of March 7th, containing Sir W. Molesworth's bitter attack on Lord Glenelg, but *not* the division. I am very happy to hear that your book is in so prosperous a state, and that you have other projects of the kind in contemplation. Pray give my best love to Emily.

Ever your most affectionate son,

C. J. F. Bunbury.

———

Graham's Town,
May 17th, 1838.

My Dear Edward,

I did not write to you on our first arrival here,

1838 because I thought you would hear all about our
journey up the country from my father, and there was
not enough of interest in it to bear repeating in
many letters. Since then we have made a ten days'
tour among the Caffers (that is to say in the country
between the Great Fish River and the Keiskamma)
—a rough journey, but uncommonly interesting.
I was present at two conferences between the
Government and the Caffer chiefs of different tribes,
one at a missionary station at the head of the Beeka
river, where the southern tribes attended,—the other
at a place called Block Drift on the Chumie river,
near the Kat Berg, where we met the chiefs of the
Gaika tribe. On each occasion there were at least
300 Caffers present. These conferences were most
picturesque and interesting scenes, the second the
most so, for it was held in the open air, in a wide
green meadow dotted with trees, where the Caffers
(most of whom were armed), seated themselves on
the grass in a semicircle around the little group of
English. The chiefs were in European dresses,
which to my taste did not become them well, suggesting too much the idea of negro footmen; but the
multitude in their native costume, and with their
native weapons and ornaments, had a singularly
picturesque and striking appearance. They are in
general tall and fine looking men, extremely well
made and active, with very animated intelligent
countenances, though their features are of the negro
type; their dress consists solely of the Ingubo, or
cloak of wild beasts' skins, which they wear in a

fashion that constantly reminds one of the toga of 1838 Roman statues,—sometimes wrapped round them, but leaving one arm at liberty, sometimes hanging in loose folds, sometimes thrown with a graceful carelessness over one shoulder, so as to leave their fine athletic figures in great part exposed, for they do not wear it from any idea of decency. They wear strings of beads round the head, necklaces of wild beasts' teeth round the neck, and brass rings round the arms and legs ; but their choicest and most elaborate decoration consists in covering their woolly hair with red ochre, not at all a more irrational practice by the way, than that of wearing hair powder. Their weapons are the assegai or javelin (which they call Umconto), and a heavy stick with a knob at the end, which they throw with great force, and with so good an aim as to bring down birds on the wing. You cannot think what a strange and wild and picturesque effect an assemblage of 300 of these fellows had, especially contrasted with the splendid uniforms of the Governor, and the English officers around him. I wished very much you could have been there and have seen it. The Caffers are certainly very clever fellows; even through the medium of an interpretation, it was impossible not to be struck with their tact and acuteness and dexterity in debate, especially their skill in evading anything that went against them, and their pertinacity in dwelling upon the weak points of the opposite case ; it would have done credit to any debater in the House of Commons. As to behaviour, I never

1838 saw a public meeting in England half so orderly as these. Both conferences ended very amicably.

The country beyond the Fish river is much finer than any I have seen within the Colony, beautifully varied with hill and dale, grass and shrubbery, with a brilliancy of verdure almost equal to England; and in many parts the broad verdant lawns and grassy slopes, dotted with trees and clumps of bush, give it quite a park-like character. The " Fish River Bush," celebrated in the Caffer war, is a very singular tract of country, very ragged and difficult not mountainous however, but a wilderness of high hills with long flat ridges and steep sides, intersected by very deep, narrow, gloomy ravines, and every-where clothed with a dense matted thicket of succulent and thorny bushes, impenetrable except to wild beasts or naked savages. This belt of wild and impracticable country is of various breadths, extending in some parts above 12 miles on each side of the river, in others not half so much ; in some places it comes within 6 or 7 miles of this town. It used to swarm with wild beasts, but the Caffers on the one side, and the officers and colonists on the other, have thinned their numbers exceedingly, and now I am told an elephant or a rhinoceros is rarely to be met with. The country immediately around Graham's Town is in general naked and ugly, but there are some pretty spots. I have been making a few attempts at landscape drawing, and get on well enough I think to recall the scenery to my own mind, if not to give an idea of it to others. You would like this place in one respect, for it is sur- 1838 rounded by open elevated plains of as fine a green turf as Newmarket Heath, where you might gallop for hours without an impediment. Thank you very much for your letter of Feb. 7th, which I have received to-day. I am delighted to hear that you take so kindly to the study of the law, and find it so interesting. I am a little amused with your complaint of the necessity of reading the newspapers ; to relish newspapers properly one ought to be as far from the fountain head, and receive them at as long intervals, as here or at Rio. *Here* indeed, in most people's minds, the interest of general politics is almost absorbed in the one engrossing topic of Caffer depredations, and Caffer treaties. Party spirit is almost as bitter and virulent here as in Ireland.

God bless you. Believe me,

Ever your most affectionate brother,

C. J. F. BUNBURY.

———

Graham's Town,
June 7th, 1838.

My Dear Father,

As I am about to return to Cape Town, I may as well write you a few lines before I start, though I am as dull as ditch-water, and have very little to say. I have resolved on returning to Cape Town with Captain Dundas of the " Melville,"*as the Governor

---

* The Hon. Sir Richard Saunders Dundas, K.C.B., born April, 1802 ; son of the second Lord Melville ; died June 3rd, 1861.

1838 does not particularly want me, and I cannot make up my mind to stay in this dull place two months more, for so long is he likely to be detained here by the accumulation of business, and of matters (both military and civil) which nobody can or will set to rights but himself. Moreover I do not like the thoughts of travelling in the winter season across the Karroos, where the cold is said to be intense. I am fully convinced that Nature never intended me for an adventurous traveller, and above all what I can least endure is cold. Besides, what I saw of the Colony (and the best part of the Colony too) in the 600 miles' journey hither from Cape Town, was by no means calculated to give a favourable idea of the whole, nor to excite any strong desire to visit the still more barren and dreary country to the north. I shall be able to spend my time more to my satisfaction at the Cape, and then in September and October, when the bad season is over and the flowers in their greatest beauty, I may make excursions to some other districts of the Colony, travelling at my own leisure, and without being dependent on another person's motions. George tells me he is writing you a long letter by this post, so I suppose he will have given you a fuller account than I can, of the laborious and unpleasant investigation in which he has been for some time engaged, touching the conduct of Captain Stockenstrom in shooting a Caffer 25 years ago. It is a subject that has made a prodigious noise in this Colony, owing to the violence of party spirit, and the peculiarly bitter hatred which the

" colonial " party (the *Graham's Town Journal* clique, 1838 &c.) entertain against Stockenstrom. For my part I must own, that I think its importance exaggerated, not only by that faction, but by Lord Glenelg too. Even were it satisfactorily proved,—which I do not think it is,—that Stockenstrom did kill a man in way which in a better state of society would be quite indefensible, I cannot see that an act committed 25 years ago, when he was a mere lad,—when his feelings were irritated by the recent death of his father,—when the killing a Caffer was an ordinary act, and not viewed by public opinion in the light of a crime, and when he saw so many examples of the same kind before him,—I repeat, I cannot see that this would be a sufficient reason for removing him from an office for which he is, in every other respect, the fittest and perhaps the only fit man. It is clear that this charge has been brought up against him at the present day solely because he has made himself so conspicuous as the friend and advocate of the Caffers, and that otherwise we should never have heard anything of it. After all, how unsatisfactory evidence must be respecting an event of 25 years ago, to which no attention was paid at the time, and which has derived all its importance from subsequent occurrences. I am in hopes that George's steady, straightforward, manly, impartial conduct, and frank conciliatory manners, are already beginning to produce a good effect in this part of the country, to calm the irritation of parties, and to restore some degree of confidence among the farmers,

T

1838 who had been deluded into the belief that the British Government was resolutely hostile to them, and that they were to be oppressed and ruined.

The good-nature with which he has received those who have come to him with complaints, and the steady determination he has shown to enforce the observance of the treaty on *both* sides, have evidently given much satisfaction; even the establishment of a new system of patrols, though it will probably not be very efficacious in preventing depredation, has done good by showing an attention to the wishes of the farmers (who have a marvellous faith in red coats), and a willingness to. do what *can* be done for their protection. However I am afraid the peace of this frontier does not rest on a very secure foundation, the Caffers, whatever Sir Benjamin may say, were by no means so thoroughly beaten in the late war as to be deterred from trying again whenever they have sufficient temptation or provocation to it, and the treaty left more than one bone of contention, especially the Hottentot settlement on the Kat River, which is a grievous nuisance in the eyes of the Caffers. The boundary too, as laid down by the last treaty, is ill-defined and puzzling. The Fish River, which forms the boundary for some considerable distance, is as bad a one as could be found, owing to the wild and intricate nature of the country along its banks (the Fish River Bush), which not only facilitates exceedingly the depredations of the Caffers in time of peace, but enables them to muster in force, unperceived, on the very frontier of the colony, when they 1838 are preparing for a *rush*. It is a strange savage country, that Fish River Bush—a wilderness of high flat-topped hills, excessively steep at the sides, with formidably deep dark narrow ravines between, and everywhere covered with one continued dense matted thicket of succulent and thorny shrubs, almost impenetrable to anything but wild beasts and Caffers.

I hope you will have received two long letters of mine from this place, one dated April 16th, the other May 16th. Our letters from England come very slowly and at very uncertain intervals; it was much better at Rio. Best love to Emily and my brothers.

> Believe me,
> Your very affectionate son,
> C. J. F. BUNBURY.

> Camp Ground,
> Near Cape Town.
> March 10th, 1839. 1839

MY DEAR EMILY,

I thank you much for your letter of the 25th of November. You will of course have heard from Cissy all about poor dear Sarah's illness.

The General was to have been married last Monday, but Mrs. Freeman fell sick, and the wedding was put off indefinitely,—however, I suppose it will take place soon, as she is convalescent. Frederic Freeman is a great addition to the liveliness of our party, having not only very high spirits, but

1839 much cleverness and abundance of agreeable conversation. Cissy has established a regular warfare with him as she did with Edward.

I am glad you agree to the truth of my remark (which of course, like other general remarks, admits of many exceptions) about political ladies, and I think you have given a very just explanation of the fact.

The truth is, that both in politics and religion, but especially in the former, ladies are too apt to have zeal without knowledge, which is pretty sure to produce intolerance; they think it unfeminine to understand or to study political questions, but not unfeminine to hate those who are on what they are told is the wrong side. This is a pity, as the influence of women is, and must be, great in politics as well as other things, and might be much more useful if they were not contented with ignorance on this subject; but I am still more averse to the fancies of Miss Martineau and the Benthamites, who would remove all distinctions between the sexes.

You seem to think that you may be in some degree liable to censure yourself: I should say certainly not, except on the subject of the negroes, on which you know I think you a little cracked.

I have very little else to tell you, as matters here jog on in the usual humdrum way. I am not yet quite certain when I shall be able to sail, but I hope in a short time, certainly in the course of the month. We are daily expecting the arrival of Charters and John, whom I shall be very glad to see again.

The decision of the Court of Inquiry held at 1839 Graham's Town last May, has been published exculpating Captain Stockenstrom from the charge of wantonly and cruelly shooting an unarmed Caffer, a decision which has very much enraged the prevailing party in the colony. I did not think Lord Brougham's second set of Political Characters so good as the first, but there was an article on Sir William Temple in the same number (by Macaulay, I suppose) which interested me much.

Have you met with " Sam Slick the Clockmaker "? there is a great deal of good fun in it, and good sense too, but one gets tired of Yankeeisms carried on through two volumes. Do you know I have become of late quite careless about botany— absolutely surfeited with it ; but I daresay the taste will revive by-and-bye.

I spend my time chiefly in reading, walking, drawing, and killing flies, which last are as trouble some here as in the south of Europe. I have done a portrait of a snake, which I think is the best drawing I ever made. Pray give my best love to my father, to whom I do not write this time, for want of matter.

Ever yours affectionately,

C. J. F. Bunbury.

P.S.—The wedding is now fixed for the day after to-morrow.

He returned from the Cape soon after the date of this letter.—F.J.B.

Batt's Hotel,

Tuesday, June 18th, 1839.

MY DEAR EMILY,

London is very full, very warm (!!!) and very
entertaining: I have seen the Exhibition (which is
very bad this year), the National Gallery, the model
of Waterloo, Macready in Henry the Fifth, Kean in
Richard the Third, and all the other sights. Henry
the Fifth is magnificently got up, and I think
Macready's acting very good, in some scenes
particularly, though not quite so superlative as some
people say; but some of the best things in the play
are left out. Fluellen is horribly caricatured.
Young Kean I do not like at all: he makes Richard
a mere ranting ruffian, instead of the *great bad man*
that he is in Shakespeare. I have seen a good many
people — Mrs. Gwyn of course, Louisa,* Miss
Freeman, Lord and Lady Euston, Lady Munro, the
Edward Romillys, and many old Cantabs, but there
are several others (Mrs. Rutherford among the rest)
whom I have called upon and not seen. I have
been at two parties, at Babbage's and Mrs. Edward
Romilly's, Henry is still in town, in better health I
think, and certainly in better spirits than I found him
at Barton. I am happy to hear from Edward that
the last accounts from the Cape are pretty good.

Pray give my love to my father, and say I will
attend to his wishes about Bauer's Ferns. I intend

* Lady Bunbury's half-sister, Miss Napier.

to return to Barton next Monday ; on Sunday, if it be 1839
fine, Edward and Henry and I are going down to
Greenwich to eat whitebait !

Ever yours very affectionately.

C. J. F. BUNBURY.

———

9, King Street,
S. James, London   1840
April 1st, 1840.

MY DEAR EMILY,

Many thanks for your agreeable letter. . . . The
moss you sent in your letter is not a Hypnum at all,
but *Sphagnum acutifolium* of Hooker, Sphagnum
palustre of Linnœus, who was probably right in considering all the different forms it assumes as varieties
of one species, though later botanists have divided
it into several species. It is one of the principal
constituents of turf bogs in all parts of Britain, and
is found also in North and South America, in Asia,
and in the Cape Colony ; in short it ranges from the
Equator to the extreme limits of vegetation within
the Arctic circle. Your violet, I suspect, is a variety
of Viola canina, which is a very Proteus of a plant,
and grows on almost all soils and situations, wet and
dry, high and low, barren and fertile. You will leave
Wales too early in the year to find many uncommon
flowering plants, but I should like to *muscologise*
there, and very likely I may do so in the course of
next month.

1840   The cold, which for some days was worse than any
we had in December or January, has departed, and
we have now mild wet weather: the *muddification* of
the streets is awful.   I called on Miss Fox yesterday
for the third time, but I am sorry to say she was not
yet well enough to see me.

I have got, according to the commission you gave
me, a very nice copy of Wilks' "Mysore" for my
father's library, in three volumes, neatly bound.   I
shall not undertake to read it here in London.   You
say Nym* is ennuyé: I suppose he sighs for—not
the flesh-pots of Egypt—but the lawns of Barton;
and remembers—not the leeks and onions that he
did eat—but the rabbits and hares that he did run
after.

In my last letter to my father I gave him all the
scraps of political news, or rather political conjec-
ture, that I could pick up, and I have learned nothing
farther.   The great corn-law battle is now raging,
but I believe nobody doubts on *which* side the victory
will be, for this time.

My best love to the dear girls.

Ever yours very affectionately,

C. J. F. BUNBURY.

Barton.
April 24th, 1840.

MY DEAR EMILY,

I arrived here on Tuesday, and have been enjoying

* A favourite dog of Lady Bunbury's.

the delicious weather, of which one certainly has 8340 more full enjoyment here than in London.

It is well too, that there should be some one here to profit by the riches of your conservatory, which is now in glorious beauty with Cactus speciosus and speciosissimus Rhododendron arborcum, Azalea Indica (3 varieties),Wisteria, Acacia armata, Tecoma australes, Zichya *alias* Kennedya coccinea (one of Henry's plants), and a variety of Geraniums, all in full bloom. The Rhododendron is very fine, with a round compact mass, as big as a cannon-ball, of deep scarlet flowers. The Wisteria is superb, and one of those in the Arboretum bids fair to equal or surpass it : I tried the other day to count the bunches of flower-buds that are formed, but it was beyond my patience. I forgot to mention the Chorozema ovatum, which is very pretty, with flowers of exactly the same colour as the Zichya coccinea. By the way, we must really take to designating some plants by *aliases* like Rogues, at the Old Bailey : thus— Tecoma australis *alias* Bignonia Pandoræ ; Zichya *alias* Kennedya ; and so on.

Out of doors vegetation appears rather backward, but there are still many beauties, among which the Ribes sanguineum is most conspicuous ; almost the whole place looks rosy with it. I had a long confab the other day with my friend Heath. Pray tell Cissy (with my love) that Whiskey* is quite a beauty, and if he did not throw back his ears would be a worthy son of his father ; he is still ridiculously

* A son of their favourite dog Nym.

1840 small, and his voice and claws are as sharp as ever. He is my companion in the evening, but I do not let him get on the chairs. The little dainty wretch will not eat bread, but is as fond of milk and of chicken-bones as his papa, and is addicted also to biting my gaiters, whereof I do not approve. Lady Blake looks a good deal better than when I saw her last, and seems in good spirits, but is yet far from being as well as might be wished. The James Blakes are in *statu quo*. Pray give my love to my father.

I forgot to mention that Peonia papaveracea is covered with buds; so is Berberis Aquifolium, but B. fascicularis does not show any. My little Sugar-bushes in the greenhouse are tolerably thriving.

Yours very affectionately,

C. J. F. BUNBURY.

JOURNAL.

1841

March 5th,   I arrived in London.

March 6th.   Went with Edward to a crowded evening party at Babbage's, where I knew hardly anybody, and was much bored. There were very few pretty women.

March 8th.   Visited the Panorama in Leicester Square. The subjects are Damascus and the Bombardment of Acre: the latter horrible from its striking appearance of reality; the view is taken from the ramparts, and the foreground consequently is full of dying and wounded men, in every imaginable 1841 position of agony, looking so frightfully real that one can almost imagine one hears them groan. The view of Damascus is beautiful. the clear bright sunny atmosphere, the warm tints of the landscape, the parched mountains, the palm groves about the city, the flat-roofed houses, the cupolas and minarets of the mosques, are admirably characteristic. In the evening went to the Haymarket, and saw Sir E. L. Bulwer's comedy of " Money," which has had a great run, having been acted 75 times this season. It was well acted, and interested me much. Some characters and scenes in it are very diverting, but on the whole it is rather of the graver kind of comedy, and is strongly marked by the general characteristics of the author. Mrs. Glover was very amusing as the widow Lady Franklin.

March 9th. Saw " Satanas and the Spirit of Beauty " at the Adelphi. It is a fantastic piece of diablerie, but very well got up, and with very pretty scenery.

March 10th. Went to a meeting of the Geological Society, where was read a long paper on the geological structure of a part of Russia, by Murchison and a French gentleman* who had accompanied him. Afterwards a lively and amusing discussion on various subjects (some of them not immediately connected with the paper), in which Fitton, Lyell, Greenough, Murchison, Lord Northampton, and others, took part. Lord Northampton is a remark-

*M. de Verneil.

1841 ably neat and fluent speaker ; Murchison not much
less so. Murchison obtained in Russia the most
satisfactory proofs of the identity of the old red
sandstone with the " Devonian " rocks ; he found
extensive tracts of red sand-stone, similiar to that of
Herefordshire, &c., containing the same shells which
are found in the limestones and slates of Devonshire,
together with the fishes characteristic of the old red
sandstone of Scotland.

After the meeting, Edward introduced me to
Mr. Featherstonhaugh, who is lately returned from
the United States. He gives a very unpleasant
account of the feeling against England in that country
and seems to think a war almost inevitable. The
best chance for peace is in the extreme poverty of
the United States, which would make it difficult for
them to support a war. The democratic party who
are going out of power, he says, have left scarcely a
dollar in the Treasury, and have endeavoured to
plunge the country into war and confusion, in order
that their successors, having their hands full, may
be prevented from examining into *their* villainies.
Mac-Leod, who he says is a worthless drunken
fellow, was not one of the party who seized the
Caroline, but boasted in his drunken orgies that he
had been ; and he chose to go into the United
States' territory out of bravado in spite of strong
remonstrances. Mr. Featherstonhaugh, however,
thinks that the object of the Yankees in his case is
rather to extort money than to put him to death.

March 12th. We have had now a whole week of the

finest possible weather, bright, calm, and warm as 1841 the end of May, so that I have been sitting with wide-open windows and no fire, and walking out at night without cloak or great coat.

March 13*th.* Went with Edward to the College of Surgeons, and spent two hours looking at the truly vast and admirable collection of skeletons. There is scarcely a quadruped, a bird, or a reptile, of which the bones are not to be seen there. It is one of the most interesting collections I ever saw. There is the complete skeleton of the famous Irish Giant, 8 feet high ; a very numerous set of human skulls, of all countries, among them some of the strange elongated skulls from the sepulchres of Titicaca, brought home by Mr. Pentland, and accurately figured in Richard's work ; several skeletons and skulls of the Orang Outan, in which the excessive length of the arms, the forward position of the head, the prominent jaws, and the huge canine teeth, show the wide interval that exists between that species and mankind ; the Chimpanzee, a shade nearer to the human race ; the massive framework of the Elephant and Hippopotamus, strikingly con-trasted with the light springy structure of the Giraffe, the Deer and the Antelopes ; the strangely formed skeletons of the Ant-eater and the Ornithorhynchus, and innumerable others. It would require repeated and lengthened visits to make oneself acquainted with the whole of this most interesting collection.

March 15*th.* This is the tenth day of uninterrupted fine weather—in March ! It is almost miraculous.

1841 To-day indeed it is warmer than it has been yet, quite summer weather. Called on Mrs. Lyell, a very pleasant as well as pretty little woman ; then went to the British Museum, and saw the new galleries of natural history, which are very rich.

The collection of birds is remarkably fine, and so well arranged that one sees all the specimens to the best possible advantage, without crowding or confusion ; but few of them are yet labelled, which is inconvenient.

There was a report yesterday that Lord Seaton (Sir John Colborne) was appointed Governor-General of India ; but it is not true, as Adair was informed at the Treasury. Lord Auckland has not resigned. Lord Seaton would have been a very good man for the purpose, but he will be wanted in America if we have war with the Yankees, as seems most likely.

Read a little book by Professor Smythe (not published) containing anecdotes of Sheridan, to whose son he (Smythe) was for some time tutor ; very entertaining. The carelessness, the procrastination, the self-indulgence of Sheridan were beyond all example, and almost beyond conception ; he was an absolute slave to his passions, his whims, and his fancies, equally incapable of self-denial in the most trifling and in the most important things ; no one could trust him or depend on him ; yet no one who approached him seems to have been able to resist his fascination, and such was the power of his genius and manner that the same persons allowed themselves to be humbugged by him over and over 1841 again.

March 26*th*. The last accounts from America are more pacific : Webster (the new Secretary of State) is well disposed towards England, and has expressed a hope of being able not only to settle the present questions in dispute without a war, but to re-knit the bonds of union between the two countries more firmly than ever. The state of New York, too, is said to have determined on removing MacLeod from Lockport to Albany, where he will have more chance of a fair trial, and be less in the power of the lawless ruffians of the frontier. It remains to be seen whether the Federal Government will be strong enough to carry out its pacific intentions, or will be forced into war by the mob. The new President's (Harrison's) speech is very lengthy, but contains not even an allusion to the affair of MacLeod.

The Lord Advocate says that Ministers have got into a difficulty about their Irish Registration Bill, having discovered that the valuations made in connection with the Poor Law, and on which the new franchise was to be founded, are not to be trusted. Will they avail themselves of this excuse to back out of the question on which they seemed to have staked their ministerial existence ? If so, they will at once throw away their Irish support, which is all they have hitherto had to rest upon.

March 29*th*. Sir W. J. Hooker, I hear, has resigned his Professorship at Glasgow, and is appointed curator of Kew Gardens in the room of Mr. Aiton.

1841 *March* 30*th.* Walked down to Chelsea, and saw Lady Wilson, and also Mrs. Vivian.—I had not seen her for very nearly five years.

March 31*st.* A pleasant evening party at the Lyells' : Owen, Whewell, young Westmacott, Dr. Fitton, &c. I had a good deal of talk with one of the Miss Horners, an agreeable girl ; also with Owen, who is a man of shy and diffident manners (somewhat like W. H. Harvey), but quite an enthusiast in his own pursuits, and ready to communicate the information he possesses so largely.

April 2*nd.* Spent an hour and a half at the British Museum, chiefly in the new Egyptian room, which contains a very rich and interesting collection of curiosities, excellently illustrating Wilkinson's beautiful book. There are scarabæi and little human figures innumerable : a great variety of seal-rings ; bead necklaces ; a very luxuriant wig (a lady's wig apparently) of dark brown hair, in profuse curls on the scalp, with an infinity of little braided tails hanging down at the sides and behind ; the wig-box (made of reeds) in which the said head-dress was kept ; a great number of little pots and vases for the toilette, mostly of alabaster, but some of richly coloured glass or porcelain ; chairs and stools, some of them made very much like camp-stools, and some resembling our cane-bottomed chairs (but the seat is in fact made of interwoven string or cord, not of slips of cane) ; combs, mostly of wood, and some of them small-toothed ; small portable metallic mirrors; &c., &c., &c. One of the most common materials

for scarabæi and other figures, beads, small pots,
&c., is a highly glazed earthenware, generally of a
turquoise blue colour, which passes on the one hand
into a bluish green, and on the other (less frequently)
into a deep blue. Where these specimens are
broken or chipped, one may see that the vitrification
is merely superficial, as well as the colour, which
was most likely produced by some preparation of
copper. Some of the little vases are made of a
nearly opaque kind of glass (as it seems) beautifully
veined or striped with dark blue, red, yellow, and
other rich colours.

April 4th. Breakfasted with Rogers the poet,
who was very good-natured and after breakfast
shewed us (Emily, Ciss and me) his pictures and
other curiosities. I never saw such a perfect gem
as his house ; everything in it bears witness to the
refined and exquisite taste of the owner ; the
pictures, the busts and casts, the Greek vases, the
illuminated manuscripts, the ornaments on the
tables and chimney pieces, all are perfect in their
kind. It would be impossible, I think, to imagine
a house giving stronger evidence of the highly wrought
polish of its possessor's mind. Rogers' pictures, etc.,
must on the whole have cost him a great deal of
money, but ten times the money would not have
made such a collection without his taste and judgment. I was particularly struck with some ancient
miniatures, illuminated frontispieces and borders of
manuscripts—some Venetian and some from the
Vatican—which are of extraordinary beauty, as re-

1841 gards both the delicate finish, and the brilliancy and
freshness of the colours. Rogers has given Hallam
the nickname of the *Boa* (*Bore*) *Contradictor*. This I
heard from Lord Lansdowne.

April 7th. Went with Emily to Chandler's nursery
to see camellias, of which he has a fine show ; but
they are not flowers that take my fancy particularly.
The Magnolia Conspicua and Berberis Aquifolium
are in full blossom and great beauty, against a wall.

April 29th. Just recovering from the small-pox,
which attacked me on the 9th, and kept me close
prisoner till the beginning of this week. I have
been reading Wood's "Journey to the Sources of the
Oxus,"—a book written in an easy and unaffected,
though not very correct, style, and containing a great
deal of curious and interesting information about
countries very little known. Lieutenant Wood seems
to have been the first European, since Marco Polo,
who has visited the source of the Oxus. This celebrated river rises in a small lake called Sir-i-kol,
on the high and dreary table-land of Pamir, in
37 degrees 27' North latitude and 73 degrees 40' East
longitude. This lofty plain, called by the Mahom-
medans the "roof of the world", is the knot or origin
of the Hindoo Koosh and several other mighty mountain chains, and on the S.E. it joins the table-land of
Thibet. The neighbourhood of the lake Sir-i-kol is
uninhabited during winter, but in summer its grassy
margin (for the snow is not perennial even at this
elevation of 15,600 feet) is frequented by the wandering Kirghiz, who pitch their tents and feed their flocks

there. "The ice upon the lake is broken up, and 1841 the hills in its neighbourhood clear of snow by the end of June." "There are no timber-yielding trees indigenous to the Hindu Koosh, in which appellation I include the range from its first rise in Pamir to its termination in Koh-i-Baba, a remarkable mountain to the north-west of Kabul. I should except the Archa, a dwarfish fir, which never equals in size its congeners of the Himalayan forests. It serves, however, for the building purposes of the natives, and is too valuable to be used as fuel. The poplar is seen by the side of most rivulets, but never in great numbers, and always in localities which indicate that man has placed it there. The same may be said of fruit-bearing trees, except the almond and pistachio-nut, which are evidently natives of the lower portions of the Hindu Koosh, and especially of the secondary ranges on the northern face of the chain. The willow of many varieties, loving a cold, moist soil, is found margining every stream, and in Durah Sir-i-kol this hardy plant was seen at an elevation of 13,000 feet. There, however, and long before that height was attained, it should be termed a bush rather than a tree. Fruit-bearing trees, of the plum species, were found at Langer Kish at 10,800 feet above the sea."

Wood gives a curious account of the Yak, or Thibet ox, called by the Kirghiz the Kashgow. "The Yak is to the inhabitants of Thibet and Pamir what the reindeer is to the Laplander. Where a man can walk a Kashgow may be ridden. Like the

1841 elephant, he possesses a wonderful knowledge of what will bear his weight. If travellers are at fault, one of these animals is driven before them, and it is said that he avoids the hidden depths and chasms with admirable sagacity. His footing is sure. . . . Other cattle require the provident care of man to subsist them through the winter. The most hardy sheep would fare but badly without human protection, but the Kashgow is left entirely to himself. He frequents the mountain slopes and their level summits. Wherever the mercury does not rise above zero, is a climate for the Yak. If the snow on the elevated flats lies too deep for him to crop the herbage, he rolls himself down the slopes and eats his way up again. . . .

" The heat of summer sends the animal to what is called the old ice, that is, to the region of eterna snow. . . The Kashgows are gregarious, and set the wolves, which here abound, at defiance." This singular animal requires so cold a climate that it cannot be kept alive in the plains, nor even at Kabul, which is 6,000 ft. above the sea-level. On the other hand, the cold of the Kirghiz country is too intense for the common cow. The two-humped or " Bactrian " camel, which Lieutenant Wood had " supposed to be a native of ' Uzbek Tartary '," (and this is the general belief) is bred, as he learned, only among the Kirghiz of Pamir and Kokan. He saw also a remarkable species of wild sheep called Kutchkar, which inhabits the snowy mountains of that savage region. " It was a noble animal, standing as high as a two-year-old colt, with a venerable beard, and two 1841 splendid curling horns, which, with the head, were so heavy as to require considerable exertion to lift them. Though in poor condition, the carcase, divested of offal, was a load for a baggage pony. The Kutchkar is gregarious, congregating in herds of several hundreds. They are of a dun colour, the skin more resembling the hide of a cow than the fleece of a sheep."

The author visited the mines of lapis-lazuli, which are situated on the North side of the Hindoo Kosh, near the head of the Kokcha river (a tributary of the Oxus), between the 36th and 37th parallels of latitude, in the province of Badakhshan. "Where the deposit of lapis-lazuli occurs the valley of the Kokcha is about 200 yards wide. On both sides the mountains are high and naked. The entrance to the mines is in the face of the mountain, on the right bank of the stream, and about 1,500 feet above its level. The formation is of black and white limestone, unstratified" (?), "though plentifully veined with lines. The summit of the mountains is rugged, and their sides destitute of soil and vegetation."

"The workmen enumerate three descriptions of *ladjword* (lapis-lazuli). These are the Neeli, or indigo colour; the Asmani, or light blue; and the Suvsi, or green. Their relative value is in the order in which I have mentioned them. The richest colours are found in the darkest rock, and the nearer the river the greater is the purity of the stone."

April 30th. Lord Morpeth's Irish Bill is smashed.

1841 A division took place last night on the 2nd clause, which is the pith, the marrow, the vital point of the whole bill,—the clause determining the franchise,—and Ministers were beaten by a majority of 11. It will be disgraceful if they cling to office after such a defeat on that which ought to be the vital point of their whole policy. If they do, it will be merely *office* that they will retain, for the *power* has clearly escaped from their hands, and the Opposition will be in effect (as far at least as the legislative power is concerned) the real rulers of the country.

On Friday, the 30th, Lord John Russell gave notice that after the Whitsuntide holidays he should move for a committee of the whole House to consider the Corn laws ; and being pressed for farther explanation, he stated that he intended to propose a moderate fixed duty on foreign corn. This announcement of course created a prodigious sensation in the House, and gave rise to an angry debate. The fact is, that it is an *ad captandum* trick, an attempt, in the desperate situation of the Ministry, to get up a popular agitation which they hope may avail them in a general election, to which they see they must be driven. It is, as the *Times* says, a mere move in the political game of chess,—an attempt to stave off check-mate by giving a counter-check to the adversary. Not that I think the plan of a moderate fixed duty on corn *in itself* a bad one, on the contrary I believe it to be the best ; but I do think the way in which these Ministers have suddenly taken it up savours much of humbug and trickery. They have

been very suddenly enlightened on the merits of this 1841
scheme. If it be so good now, why was it not equally
good last year, or the year before ? But I have no
doubt they will be beaten on this question also, and
probably by a larger majority than in the case of the
Irish Bill. They will have the whole body of county
members against them. Augustus O'Brien tells me
that the Conservatives do not intend to allow Ministers
a respite on this question, but that Peel will move, in
the course of this very week, a resolution to the
effect that it is expedient to make provision for the
public service, etc., without interfering with the
present duties on corn. Thereupon a division will
ensue, and the Ministry, I expect, will be soundly
beaten.

May 6th. Visited the Royal Academy exhibition
a second time. There are several satisfactory pictures, though the absence of Edwin Landseer is a
sad loss. Stanfield's Castle of Ischia is admirable ;
Roberts' Temple of Denderah is a very striking
picture, and gives a fine idea of the massiveness and
vastness of those stupendous buildings, in which
even the profusion of painting, and the rich colours
bestowed on the sculptured figures, do not injure
the grandeur of the general effect. There is a beautiful picture of the ruins of Baalbek, by the same
artist, and also a view of Jerusalem, without much
pictorial effect, but apparently of great local truth.
Christ's Lamentation over Jerusalem, by Eastlake,
seems to me one of the very best Scriptural pictures
of the English School, which indeed is not very high

1841 praise.   Cooper has a spirited painting of a cavalry combat, the fight at Cropredy Bridge.   Turner is less extravagant in his colouring than usual, particularly in two views of Venice, which have really some resemblance to nature, though there is still too much scarlet and yellow in them.   Martin has a striking picture of Pandemonium, grand and gloomy, with a fine wild effect of lurid volcanic light ; Etty, two of his favourite subjects of naked women, painted with his warm voluptuous glow of colouring.   There is a clever and well-painted, but not agreeable, picture by Herbert, Venetian brides carried off by pirates ; and two very pretty Italian subjects by Collins—a name I am not familiar with,—particularly a view from the " Caves of Ulysses " at Sorrento.   Jones has some pretty sepia drawings, and he is wise to keep to them ; but he has not made so much as he might of Lord Wellington at the Battle of the Pyrenees.   In the sculpture room, one of the most attractive things is a group of two boys feeding a little bird, very natural, easy, and graceful.

The corn-growers are not the only powerful interest that the Ministry, in their despair, have determined to brave.   In bringing forward the Budget on the 30th, the Chancellor of the Exchequer announced the intention of reducing the duty on foreign sugar, so as to permit its importation.

This it seems is to be the first point of attack, and the debate on it is to come on to-morrow (the 7th), Lord Sandon leading the attack, which is very artfully, if not very honestly, grounded on the objection to admitting slave-grown sugar, and thereby 1841 encouraging the employment of slave-labour.

The Tories have cleverly chosen their ground so as to engage the vehement abolitionists in thei assault on the Ministerial position; and they wil most probably succeed. There is nevertheless much to be said in favour of the admission of Brazilian sugar at a reasonable rate, for our commercial treaty with Brazil will expire before long, and that country, which now takes a very large quantity of our manufactures, will probably give the preference to the German manufacturers if we refuse to take her sugar in return. We have already in our hands a great part of the carrying trade of Brazil, and should most likely have the whole if we dealt with her on fair and equal terms.

May 9th. Lord John Russell stated on Friday (the 7th), what was the rate of duty he intended to propose on foreign corn,—8s. a qr. on wheat, 5s. on rye, and so on; not high enough to satisfy the farmers, I take it.

The same night, the great debate on the sugar question, which is expected to decide the fate of the Ministry, was opened by a good speech from Lord John,—some say the best he has ever made. Lord Sandon moved his amendment, which was seconded by Mr. Hogg; Mr. Hawes supported the ministerial proposition, and Dr. Lushington (foolishly I think) opposed it. The debate was adjourned to Monday (to-morrow) and is expected to last Tuesday night also.

1841   *May* 10*th*.   Visited the Water-colour Exhibition, in which there are many very pretty things ; but it has has a great sameness in it too ; year after year it is always the same thing, subjects of the same kinds treated by the same artists, so that one knows as well as possible what to expect.   Copley Fielding is, as usual, far above all the rest, and his View from Bolton Park is one of the most beautiful even of his drawings.   Gastineau has some admirable things, particularly " Dunluce Castle," a magnificent scene of precipices and ruins and wild stormy sky and raging sea.   There is a large drawing by Lewis of the Piazza di S. Pietro on Easter Day, very attractive ; but his Spanish recollections seem to hang about him too much ; he has made his Romans too like his Spaniards, and has given to the women in particular the sly half-sleepy Spanish eye, which is out of character.

May 14*th*.   Went into the British Museum for a short time.   The whole circuit of the Natural History rooms is now open, and a magnificent series it is. In the mineral room I was much struck by a suite of uncommonly fine specimens of opalized wood from Van Dieman's Land, some of them very large masses, and exhibiting in the most beautiful manner all the gradation from wood in appearance almost unaltered, to the fine semi-transparent brown opal with a perfect conchoidal fracture.   I noticed also some very large mamillated masses of native arsenic, from Andreasberg in the Hartz.

May 15*th*.   Met Lord Charles FitzRoy in St.

James' Park, and had a long talk with him. He is 1841 very eager in behalf of free trade, and has published an address to the electors of Bury, declaring himself a decided advocate of it, and resting his claim to their support on this ground. I am afraid this will not tend to promote his cause there, but it is manly and straightforward, and therefore much the best course.

He is to be opposed, it seems, by Horace Twiss, who has gone down thither to convass. Lord Charles is more sanguine than I can feel about the effect of this free-trade question on the elections ; he thinks that the Ministers may even gain a majority, which appears to me very unlikely, considering the superior organisation and activity of the Tories, their attention to the business of registration, and lastly, that if the proposed alteration of the Corn Law gains the Ministers some votes in the boroughs, it will deprive them of what little support they have at present in the counties. He says the debate (which has now lasted six nights, and is again adjourned), is pur- posely lengthened out by the Ministerial party, in order to give time for public meetings, and for the opinion of the country to be pronounced upon it.

I afterwards met Henry Arundel, who of course is vehement against the new measures. He tells me that the Tory estimate of the majority they are to have on the approaching division varies from 25 to 40. Visited the Adelaide gallery, and noticed, among other things, some rich specimens of sul- phuret of silver, from a mine in the island of Sark.

1841   *May* 16*th.* Walked in St. James' Park, which is
very pretty and enjoyable. On a fine Sunday at this
time of year, with the trees in the full beauty of their
young foliage, the shrubberies bright with the blossoms of the lilac, laburnum and hawthorn, the grass
in its liveliest verdure, and the swarms of people in
their best dresses enjoying the holiday and the sunshine, it is a charming scene. The collection of
curious ducks and geese on the lake is very
numerous, comprising most of the British and
several foreign species; they nearly all are very tame
and come eagerly to receive bread from the visitors.
It is said that many of the young ones are eaten by
the fish. Charles the Second had a collection of
"curious fowle" in this same place, repeatedly
mentioned by Pepys.

The report seems to gain ground, that the Ministry
if defeated on the sugar question (as nobody doubts
that they will be), will resign immediately, without
dissolving Parliament. Edward heard at Lady
Colborne's that Lord Melbourne, Lord Lansdowne,
Lord Normanby, and one other of them, were
determined on this step. For my part, I think they
ought to dissolve, and give the country a fair chance
of expressing its opinion on the very important
question of free trade. The Queen gave a ball
onFriday (the 14th), and some say *that* is the
reason the division did not take place that
night.

May 18*th.* This "monster debate" has been
again adjourned, after lasting seven nights! Last

night most of the speakers were obscure men, whom 1841 one has seldom if ever heard of.

May 19th. The long expected division took place this morning at 3 o'clock, and Lord Sandon's amendment was carried against the Ministers by a majority of 36, in a house of 598. Sir R. Peel's and Lord Palmerston's speeches, on this last night of the debate, were excellent. The Ministry have not resigned, and do not seem to intend to do so without a dissolution, and as far as I can understand Lord John Russell's announcement, it seems that they mean to encounter another defeat, on the Corn question, before they dissolve. Perhaps ·they wish to allow a week or two longer for the anti-corn law agitation to spread and develop itself.

May 20th. Went to an evening party at the Horners', which was very pleasant, *barring* the music. I was introduced to the famous Robert Brown, the greatest of botanists, whom I found pleasanter and more communicative than I had been led to expect.

May 21st. From London to Hastings, a pleasant journey of eight hours, through a very pretty country for the most part. There are very fine beeches on the chalk hills near Sevenoaks, and a few miles on the London side of that town, the road, winding along the escarpment of the great chalk range, commands a very extensive and striking view of a considerable part of Kent and Sussex. The Chancellor of the Exchequer last night gave notice that on Monday the 24th he should move the renewal of the ordinary annual sugar duties, which would otherwise

1841 expire; and Lord John Russell announced that he
should bring forward the corn question on the 4th of
June; so it seems there will be neither resignation
nor dissolution before that. A dissolution, however,
early in June, seems to be now looked upon as
certain, and great preparations are making for the
contest.

May 25th. I find Hastings a pleasant place, much
superior at any rate to any of the watering places on
our coast. The greatest part of the town is huddled
together in a deep valley opening to the sea between
two steep green hills, which, running down from the
high lands of the interior, terminate on the coast in
bold rocky escarpments; these are called the East
and West Hill; on the top of the latter are the ruins
of the old castle. The *fine* part of the town consists
of a row of houses along the top of the beach, immediately under the cliff on which the castle stands.
The country inland from Hastings is hilly, very
pleasant and pretty, with much variety of surface,—
elevated downs, deep fertile valleys, narrow, winding
lanes and hollow ways between shady banks and
much copsewood. The hedge-banks and lane-sides
are beautifully clothed with ferns, in great luxuriance.

May 26th. The warfare of politics seems to be
going on with spirit, and the Conservatives are following up their victory. On Monday Sir Robert
Peel gave notice of a motion of want of confidence in
the Ministry, which will be brought on to-morrow
(Thursday), and will no doubt give rise to a very
sharp and protracted debate; I expect it will be

carried, though it may be a near thing. The renewal 1841 of the ordinary sugar duties was passed without opposition. Ministers have abandoned the Poor Law bill, seeing no prospect in the present state of public business (and of public feeling, it might be added), of carrying it through in this session. So, good-bye to the mighty Commissioners.

June 5th. Return to London, and find that Sir Robert Peel's motion (want of confidence in Ministers) has been carried, after several nights' debate, by a majority of *one*, the numbers being 312 and 313. It is rumoured that Parliament will be dissolved next week.

June 8th. Astley's Theatre burnt to the ground early this morning. I had been in it the last evening, only a few hours before the catastrophe.

June 9th. Return home.

June 10th. Barton is now in great beauty, though the ravages of the late severe weather are in some instances conspicuous. The Arbutus and Bay are almost all killed, and the common Laurel much injured ; the common Myrtle, though against a south wall, and protected by mats through the winter, has been killed to the ground, but is shooting out again from the root ; the Ilex and Cypress are but slightly hurt, and scarcely any of the North American plants have suffered at all ; Araucaria imbricata, in the open ground, but carefully covered up during the frost, is uninjured ; the Cork tree not quite killed. The last winter seems on the whole to have been more destructive than that of 1837-8. It is observable that the

1841 Pontic Rhododendron is uninjured by the utmost
severity of our winters, while the common Laurel, a
native of the same country, often suffers severely.

June 16th. Lord Charles FitzRoy dined here. He
has been canvassing the electors of Bury in conjunc-
tion with Mr. Alston, against Lord Jermyn and the
old Tory hack Horace Twiss. The contest will be a
very hard one, but my belief is that neither of the
new candidates will come in.

June 20th. Colonel Wilks has some curious observations on the real pretensions of the rival candidates for the thrones of the Deccan and the
Carnatic, whose claims, supported respectively by the
French and English, led to the first important war in
India between those nations, and laid the foundation
of our Indian empire. It is to be premised that
Nazir Jung and Muzuffer Jung (who were uncle and
nephew) were the candidates for the sovereignty (or
nominally the viceroyalty) of Deccan; Mohammed
Ali and Chunda Saheb for the subordinate government of Arcot.

June 22nd. Called on Mr. Richards, the clergyman of Stowlangtoft, who has a considerable collection of miscellaneous curiosities. He shewed me a
singular Druidical relic, a remarkably exact imitation
of an *eye* in coloured glass, set in a triangular-shaped
piece of a peculiar-looking brownish stone or composition. This was dug up at Thetford, and with it
were found some balls of agate, polished, and pierced
for stringing. He told me that the Fly Ochis had
been shown to him that year by Lord Charles

Hervey, in a field at Ickworth, in uncommon abun- 1841 dance,—that there were some hundred plants of it in the field. Miss Rickards finds Medicago Maculata and Veronica Polita in Stowlangtoft parish. Here also I saw a fine specimen of the Crossbill, shot not long ago in the neighbourhood, in full red plumage, which I believe is rather unusual.

June 23rd. Parliament was dissolved yesterday.

June 28th. Now begin the elections, and many of the most important will be decided in this week. The nomination at Bury took place to-day, and I had intended to be there, but the morning proved extremely wet, and I did not go. The nomination for the City of London was also to be to-day, and the election will take place to-morrow: that is always one of the most important elections, and peculiar interest is attached to it this time in consequence of Lord John Russell being one of the candidates.

June 29th. The Bury election has terminated, as I expected, in the return of the two old members, but after a very sharp contest. At the close of the poll, the numbers were: for Jermyn, 341; Fitzroy, 310; Twiss, 298; Alston, 256. There were several windows broken and some heads.

June 30th. The city of London has returned two Liberals and two Conservatives—Lord John at the head of the poll, Wood next, Masterman third, and Lyall fourth. Each of these polled above 6,000 votes. I daresay the representation of many other places will be similarly divided, and in fact it seems

1841 but fair and reasonable that, where the two parties
are pretty equally matched, each should be represented. The preceding sentences were written this
morning, and the account of the London election
was taken from the *Sun* newspaper. But I find
there is a strange discrepancy in the accounts of it ;
according to the *Standard*, Lyall is at the head of
the poll, Wood second, Attwood third, and
Masterman fourth—Lord John being thrown out.
The true state of the case will not be known till the
official return is made. In the meantime it is
certain that the Conservatives have very much the
upper hand as far as the elections have gone
hitherto ; of those borough seats for which there has
been no contest, they have more than two thirds,
and at Cambridge, Harwich, Reading, and St.
Albans, they have been completely victorious. I
now begin to think that their anticipations, of a
majority of 100 at least, will be verified.

July 1st. The last accounts of the progress of the
general election are more satisfactory than those
received yesterday. The Tories are still ahead, but
the Whigs and their allies have gained some important victories : the most remarkable is at
Nottingham, where, but a few months ago, Walter
came in triumphantly, but where he this time has
not ventured even to go to the poll, such was the
superiority of Hobhouse and Larpent. At Bath,
my old college acquaintance Lord Duncan (a Whig)
has come in together with the ultra-Radical
Roebuck ; at Coventry, Ellice and Williams ; at

Derby, Strutt and Ponsonby; at Northampton, 1841
Vernon Smith and Raikes Currie; all Whigs. The
Tories on the other hand have been successful at
Bedford, Lincoln, Pontefract, Rochester, and some
other places. St. Albans has not returned two
Tories, as was asserted, but one of each party. Two
of the Whig Ministers, Palmerston and Baring, have
come in without opposition.

Colonel Wilks (whose " History of Mysore " I am
reading with considerable interest) combats the
notion entertained by many, that the use of fire-arms
was known in India earlier than in Europe.

July 2nd. The real result of the London election,
as announced by the Sheriffs, is different from the
statements of *both* the Committees. Masterman is
at the head of the poll, Wood second, Lyall third,
and Lord John Russell fourth; the members thus
being two of each party. In Westminster the Tories
have gained a singular and unexpected triumph,
their candidate, Captain Rous,* having come in at the
head of the poll, though he had been in the field but a
few days; the rejected candidate is General Evans,†
who is no great loss. At Liverpool also the
Conservatives have won hollow; at Manchester, on
the other hand, Phillips and Gibson have been
successful, and at Yarmouth the two Liberals have
come in by a majority of nearly two to one. The
whole number of members hitherto elected, according to the *Chronicle,* is 92 Whigs or Radicals, and

* Brother of Lord Stradbroke.
† Afterwards Sir Lacy Evans.

1841 104 Conservatives; but the latter will gain many more in the counties. The excessive drought which lasted through the greater part of May and June has been succeeded by equally excessive rains.

July 3rd. The elections are going on better than seemed likely from the beginning: Colonel Fox has come in for the Tower Hamlets, at the head of the poll, and Clay with him; Sir B. Hall and Commodore Napier have been successful at Marylebone; Ipswich has returned two Reformers, so has Sheffield, and so has Brighton. The Tories on the other hand have several triumphs to boast of. On the whole, it is asserted, in those boroughs which have changed their representatives in the present election, the Whig-Radicals have gained 27 seats, and the Conservatives 37.

I read the 5th Satire of Juvenal, which is by no means one of his best, but is useful to read for the trial and improvement of one's scholarship, being remarkably full of hard words and obscure allusions. It is the satire which describes the miserable life of a poor dependent on a great man, and especially the insulting distinctions made at table between the fare and treatment of the great man himself and those of his humble guests. If this author may be trusted, the Roman aristocracy ventured on higher flights of arrogance and overbearing insolence than ours ever have attempted. I think Pliny the younger also speaks with censure of those offensive distinctions at table.

July 6th. The two Tory members for West Suffolk,

Waddington and Rushbrooke, were re-elected with- 1841 out opposition.

July 7th. According to the *Standard*, the result of the whole of the borough elections for England and Wales is that, as compared with the last parliament, the Tories have lost 31 seats, and the Reformers 39—giving to the former, on the balance, a gain of 8. They brag of this, but for my part I expected that we should lose more.

July 16th. The English and Welsh elections are complete. The boroughs have returned 176 Liberal members and 166 Conservatives; the counties, 22 Liberals and 137 Conservatives.

July 17th. Wisteria Sinensis, which usually blossoms in March and April, is now coming into profuse bloom in our conservatory; this second crop ,of flowers having been encouraged by cutting off the first before they expanded, early in March.

I have several times lately noticed here the Lesser or Red-backed Butcher-bird, a very pretty little bird, which I had not observed before. Its note sounds like a repetition of the words ' chat-chat-chat,' in a short, quick, harsh tone; it will sit on a branch for a pretty long time together, repeating these sounds without variation, and at the same time continually jerking its tail up and down with a quick motion

July 20th. Read Colonel Wilks' detailed and interesting narrative ("History of Mysore", chap. 22) of Hyder Ali's famous invasion of the Carnatic in 1780, of the destruction of Colonel Baillie's detachment, and of Sir Eyre Coote's campaign against

1841 Hyder. The almost inconceivable negligence, stupidity, and imbecility of the Madras government of that day form a striking exception to the generally brilliant career of the British nation in India. It seems almost a miracle that the English were not altogether rooted out of the South of India by their great and terrible enemy, while their councils were directed by men so incapable, and while they were at the same time cursed with the alliance of that most abandoned villain Mahommed Ali, the Nabob of Arcot.

Hyder Ali was a very extraordinary character; one of the most striking examples in history of the sort of character that is formed by the most vigorous and varied natural powers of mind, unadorned by cultivation, unchecked by any particle of justice, faith, or mercy, or by any moral restraint whatsoever. On the whole he strikingly resembled some of the ablest Italian tyrants of the middle ages, though his abilities and his wickedness were displayed in a wider field, and were perhaps carried to a higher pitch even than theirs. No man that ever lived, probably could surpass him in craft and dissimulation, and it is no wonder that these were his favourite weapons; and although he had formed a very great and powerful military force, and brought all the establishments connected with it to a high degree of perfection, he seems to have always preferred gaining his ends by fraud, except when his superiority was quite overwhelming. The extraordinary perfection of his system of police and 'espionage,' both for domestic and foreign affairs, was one 1841 of the main sources of his great success. He was subject to violent fits of passion, snd was lascivious in the extreme, but never allowed either his anger or his lust to interfere for a moment with the prosecution of his schemes of avarice and ambition.

July 25th. To London.

July 27th. From London to Bridgwater, by the Great Western Railway, in less than 6 hours ; and thence by coach to Exeter, 5 hours. The arrangements on the railway seem to me excellent, and the speed is really wonderful ; the first-class carriages are very comfortable, and I did not find the motion annoying, in general, though it varies considerably, from what cause I cannot tell. The passing through the tunnels, particularly one enormously long tunnel between Bath and Bristol, is not very pleasant. Altogether it is an admirable way of travelling when one's only object is to pass rapidly from place to place, but not if one wishes to see much of the country.

July 28th. Travelled post from Exeter to Devonport, by Chudleigh, Ashburton, and Ivy Bridge, through a very pretty, pleasant, and varied country. Crossed a part of Haldon hills, high wild moorlands glowing with the bright tints of the furze and heath flowers, and making a fine contrast to the beautifully rich and verdant lower country, of which they command an extensive view. Dined with Admiral Sir Graham Moore, whose house is situated in the pleasantest part of Devonport, on the height

1841 called Mount Wise, overlooking the Sound, and directly opposite to the beautiful woods and green slopes of Mount Edgcumbe.

July 29th. Went into Plymouth in the morning, got myself enrolled as a member of the British Association, and in the evening attended the first general meeting, held at the Town Hall, Devonport. It was very much crowded. Whewell, the President, made a long speech, in part good, but pompous; Lord Northampton spoke well, and so did Sir Thomas Acland. The whole was a sort of parade affair, of no great interest.

July 30th. Attended the Natural History section (D) of the Association, from 11 o'clock to near half past two, and heard several papers read. A paper, by Captain Widdrington, on the natural history of the eel, led to a long conversation, in which many curious particulars were mentioned by Mr. Couch (the famous Cornish ichthyologist), and others, relating to the eel and various other fishes; several instances were mentioned of the sudden and extensive mortality that at times occurs in particular species of fish, and of which the causes are not known. Mr. Couch noticed also a remarkable case in which the roe of the whiting proved poisonous, and another member spoke of the very deadly properties possessed at times by certain fishes of the West Indies, the poison of which is in some instances so violent as to cause death in less than ten minutes, and with terrible convulsions.

Next came a very short but curious paper by Mr.

Ball, who having put the head of a porpoise, in a 1841 fresh state (not putrid), into one of the closed fern-houses on Mr Ward's plan (the dimensions of which were 30 cubic feet), found that in a short time all the plants contained therein were more or less withered, seemingly by the exhalations from this animal matter. The room was very well filled. Among the company was a Parsee from Bombay, a picturesque figure, with a very clever countenance.

July 31*st.* I accompanied Sir Graham and Lady Moore, and their party, in a very pleasant boating excursion up the Tavy. Above its junction with the Tamar, this river soon assumes a very picturesque character, winding in graceful curves between high and steep hills beautifully clothed with wood. Warleigh woods are the first we come to in ascending the river; then Maristow, a large white house, situated on a broad green lawn which descends with a rapid slope to the water, between thick woods; and opposite to it, a very fine bank of wood. We landed a little above Maristow, and walked up the hill to Buckland Abbey, a building not remarkable in outward appearance, but noted as the residence and property of Sir Francis Drake, whose sword, drum and Bible are preserved here. There are very fine tulip trees in the garden, and long avenues of noble beeches and limes near the house; and from various points we caught charming glimpses of the winding course of the Tavy, between its wooded hills, and of the broad lake-like expanse of water at its junction with the Tamar. From hence we

1841 descended to the river side, and walked back through pleasant woods and meadows to the place where the boats were' waiting for us. The hanging woods on the right bank are of great extent and beauty, and there are some magnificent trees, chiefly oak and ash, in the meadows, on the edge of the water. We were out about six and a half hours.

August 1st. In a walk among the lanes within a few miles of Devonport, I observed (besides such plants as occur everywhere), Geranium, Columbinum, G. rotundifolium, Pimpinella magna, in great plenty, and Malva moschata.

August 2nd. Attended the Natural History section again, and heard some things worth remembering. The discussions to which the several papers gave rise were much more interesting than the papers themselves. Mr. Couch mentioned several facts relative to the birds of Cornwall, and their migrations; among others, that the swift is the most constant and regular of birds in the time of its appearance in this country, and always arrives in flocks, while the common swallow and the martin make their appearance singly. The stormy petrel is a common bird on the coast of Cornwall in bad weather, and is driven far inland when it blows hard. Mr. Couch thinks that this bird does not possess the power of contending or making head against a gale, but is always driven before it, and that its supposed property of being the harbinger of storms is owing to this. The golden-crested wren, which is a constant resident in England, is migratory in

Belgium. A tedious, ill-written, inconclusive paper 1841
on the analogies of animal and vegetable life,
gave rise to a conversation of some interest, in which
Professor Henslow took a prominent part; it turned
upon various points of vegetable physiology, and
especially on the much-disputed question of the
excretions of plants, with respect to which the
general opinion seemed to be that much more
information was required before the point could be
settled. A very curious fact was mentioned,
relating to a new method of dyeing wood while
growing, by making incisions in the trunk and
infusing various coloured saline solutions ; the wood
becomes beautifully coloured, and the tree itself lives
on, but these foreign matters, being excreted by the
roots, poison the neighbouring trees. There seems
to be no doubt that matters which do not properly
form part of the food of a plant, if taken into its
system in any way, are excreted by the roots ; but
the disputed point is whether this takes place with
the natural food of the plant.

In the afternoon I went to the dock-yard to see
the launching of an 80 gun ship, the " Hindostan,"—a
noble sight, but one which I cannot attempt to
describe. I was indebted to Lady Moore's kindness
for a ticket which procured me a most excellent
place.

August 3rd. Again to Section D. Professor Owen
gave us some observations on the Thylacinus, one of
the largest and rarest of the marsupial quadrupeds,
peculiar to Van Dieman's Land, where it is rapidly

1841 diminishing in numbers through the active warfare waged against it by the colonists, and is likely soon to be quite extinct.   It is a great destroyer of sheep. and resembles the wolf in its general habits, but shews much less cunning, and indeed a degree of stupidity. A perfect skeleton of it has now for the first time been brought to Europe, and Owen stated several peculiarities of its organization, in which it may be considered as a lower grade or type than the non-marsupial carnivora.   He also noticed the very general prevalence of the marsupial structure in the mammals of Australia, and conjectured that this might be connected with the exceeding drought of that climate, which made it frequently necessary for the animals to make long nocturnal journeys in search of water, when the young might be liable to perish if the dam had not the power of carrying them with her.   But is not the climate of Southern Africa as dry as that of Australia ?   In the young animals of this marsupial tribe, it appears, the pouch exists (in a more or less rudimentary state) in the males as well as the females, and the traces of it are unusually distinct in the male Thylacinus. A Mr. Bellamy read a paper on the mammals of South Devon, which was chiefly remarkable for his including among them the *mermaid*, and giving a detailed description of one, from the statements of some sailors who had actually caught one, but from a superstitious fear had let it go again.   An amusing discussion took place on this topic ; Owen shewed considerable incredulity, and several instances were

mentioned of fictitious mermaids which had been 1841
imposed on the public. It is certainly unlucky that
the mermaid, like the unicorn, appears only to the
uneducated, and seems always to shun the eye of
science. There was also a good deal of conversation
respecting the changes of colour in the hair and
feathers of animals: Mr Couch mentioned that the
stoat, weasel, and hare, are often found white in
Cornwall, where this change cannot be attributed to
cold, as it is not confined even to the winter season,
much less to a severe winter. Colonel Hamilton
Smith mentioned a Canadian hare, which being
caught when in its white dress, in October, and put
into a cage to be brought to England, resumed its
brown colour before the end of the voyage. A paper
was then read on the subject of some skulls brought
from Bolivia, which were of the same extraordinary
elongated form as the Titicacan skulls in the
Museum of the College of Surgeons. The circumstances under which they were found were detailed,
and two of the skulls (those of children) were
exhibited, together with the wrappers and various
other articles found with them. The author of the
paper inclined to the belief that this singular form of
the skull was natural, but Owen thought otherwise,
and gave us an admirable lecture on the means by
which it might have been produced, the traces of
such pressure which were visible on it, and various
other points connected with the subject. Colonel
Hamilton Smith afterwards read a curious paper on
certain gigantic cuttle-fish, having collected with

1841 great care and industry all the recorded instances of their occurrence in ancient and modern times. Altogether, this meeting of the section (the last) was of unusual interest. Weather excessively bad these two days.

August 4th. The Devonport regatta took place this day; the weather fortunately very fine. The steep green face of Mount Wise was covered with spectators of all classes, in their best dresses; the water alive with pleasure-boats; the sun shining bright, and the whole scene exceedingly gay and animated. In the evening I attended the last general meeting of the Association, heard a great deal of pompous complimentary speechifying, and grew very tired of it.

August 5th. Left Devonport, and travelled post, by Callington and Liskeard, to Glynn House, Sir Hussey Vivian's, near Bodmin. Saltish, where I crossed the water, is a dirty little fishing town, but rather picturesquely situated on the side of a hill rising steeply from the water. Thence to Liskeard the way is through deep narrow lanes between high banks, which allow little to be seen of the country. The new road from Liskeard to Bodmin runs along the pretty wooded valley to the Fowey river, a rapid and brawling stream like those in the north of Devonshire.

August 6th. Glynn House is pleasantly situated on the slope of a hill facing to the south, above the Fowey river, with extensive woods on both sides and opposite to it. The valley is narrow, the hills high

and steep, well wooded on their slopes towards the 1841
river, open and heathy at their tops, not shewing
much rock. The vegetation, both of the woods and
moors, is very much the same as that of the
neighbourhood of Linton in Devonshire, to which
indeed the general character of the country is very
similar. I found however two plants which I had
not before seen alive : Bartsia viscosa, in a moist
place beside the old road to Bodmin ; and Sibthorpia
Europæa in the woods of Glynn. Ferns, mosses
and lichens are in great profusion, and of luxuriant
growth. The conservatory at Glynn contains a
number of good plants, in uncommonly fine condition,
such as Datura (or Brugmansia) arborea, D.
sanguinea, Siphocampylus bicolor, &c ; the Sollya,
trained against a trellis, is nearly 20 feet high, and
Maurandya Barclayana, in the same situation, has a
strong woody stem as thick as my thumb. The
Fuchsias are of exceeding beauty. The hay harvest
here is even later than with us in Suffolk, and the
corn is cruelly beaten down by the late stormy
weather.

August 7th. Walked down the valley of the Fowey
river from Glynn to Restormel Castle, a very
pleasant excursion, going down the right bank of the
river and returning by the other. The valley is
beautifully green, with fine woods along the hill-sides
and in many places down to the brink of the rapid
stream. Restormel Castle, about the history of
which I could not gather any precise information
except that it is believed to be older than the

1841 Norman times, stands in the midst of a wood, at some height on the hill-side, overlooking a long reach of the valley and the pretty little town of Lostwittiel. The body of the castle, a circular battlemented tower of large dimensions, surrounded by a deep ditch, remains in pretty good preservation, but so completely overgrown with ivy that the architecture can hardly be made out. The gateway is a good piece of ruin. This castle belongs to the Duchy of Cornwall. I found Gymnostomum vividissimum with abundance of its thecœ on trees hard by the ruins, and several curious Lichens in the woods nearer the river, the Sibthorpia on a dripping bank at Respryn.

August 8th. Went to church at Cardinham (or Caerdinham) heard an excellent sermon, and returned (in company with Sir Hussey and Lady Vivian, Baron Rolfe and others,) by a circuitous and very bad road through a wooded valley which is tributary to that of Glynn. The extent of wood in these parts is very considerable, and the scenery often reminds me of that near Linton. The prevailing characteristics of this part of Cornwall, as of the north of Devonshire, seem to be high bleak moors, with very pretty romantic, wooded valleys deeply excavated between them.

Sir Hussey Vivian's only child by his present wife, a very pretty little girl, has a name which I never met with before in real life—Lalagé.

In front of the house at Glynn are four French light guns, three-pounders, taken in the Pyrenees.

The garden is in the Italian style, and there is an 1841 excellent copy of the Medicean Venus in a little *tribune* or cabinet adjoining the conservatory.

August 9th. I left my hospitable friends at Glynn, and proceeded to Fowey, through very narrow, very crooked, and in some places very steep lanes, with high banks forbidding any general view of the country. The descent into Fowey is about the steepest bit of road I have seen in England. This is a vile, dirty little fishing town, but situated on a fine estuary (a good deal like Falmouth harbour, though on a smaller scale), which opens to the sea between two bold rocky headlands. The indentations of the coast are remarkably picturesque; the rocks not very high but exceedingly abrupt, sharp, and jagged; and among them are many quiet little coves and nooks, very pleasing to the imagination. The little pools left by the receding tide amidst the rocks are beautifully lined with a variety of Fuci, Actiniæ, and Corallines. The Crithmum grows in many places on the cliffs, and also the Brassica oleracea, I believe, but in inaccessible places.

The inn at Fowey looked at first sight very unpromising, but turned out more comfortable than I expected. The town is built on such steep ground that the houses appear to stand one on the top of another. Opposite to it, in a somewhat similar situation, is the village of Polruan, and some fragments of old towers, perched on the rocky headlands, improve the effect of the scenery.

August 10th. From Fowey to St. Austell, a town

1841 of some consequence, but ill-built. Its neighbourhood abounds with tin mines, but the deplorable weather prevented me from seeing anything.

August 11*th*. To Carclew, passing through the once noted borough of Grampound, the village of Probus (which has a very handsome church tower), and the important town of Truro.

August 12*th to* 15*th*. At Carclew, which is a very comfortable house, and in every way one of the most agreeable I have been in. The gardens are very fine, the warmth and moisture of the climate producing a luxuriance of vegetation which I have seen nowhere else in England.

The trees and shrubs of N. America and of Nepaul are here in surprising beauty : the Rhododendron arboreum flourishing in the open ground, and growing to a really tree-like size ; R. maximum equalling in height our largest laurels ; Magnolia grandiflora, standard, from 15 to 20 feet high ; Pinus Morinda and Webbiana growing most vigorously ; the Tulip tree attaining really gigantic dimensions. The Aristotelia, from Chili, and a Grevillea, are also quite hardy here, as well as the Fuchsias.

The flower garden is very prettily arranged in successive terraces, which have an excellent general effect, and are glowing with brilliant blossoms. The houses also are richly stocked, especially that appropriated to the orchideous plants, few of which however are at present in flower. In another of the hothouses, I observed Limnocharis Humboldtii in great perfection, and the curious Nepenthes,

The park is extensive and rather pretty, with con- 1841 siderable variety of surface, and some good trees, in particular Levant and Luccombe oaks of unusual size. The sessile-fruited oak appears to be here more common than the pedunculated. The neighbouring country is more open in general than the eastern part of the county, and though the vegetation is so luxuriant in the sheltered parts, yet on the higher parts trees do not grow very well, not owing to the cold, but to the violence of the prevailing winds, and the salt vapours which they bring with them from the sea. The surface of the country is very hilly, but the hills are tame and unpicturesque, and are on a smaller scale, and generally less steep, than in the more eastern parts. The harvest is later than in other parts of the south of England, owing to the excessive quantity of rain. Among the native plants, Erica ciliaris is the great treasure of this neighbourhood ; it grows in abundance among furze on dry banks and rough ground, in a blackish heath soil (not boggy), on the north western skirt of Sir Charles Lemon's demesne, towards the little village of Perran. Near the same place I gathered Bartsia viscosa in plenty, Sibthorpia Europea, Pinguicula Lusitanica Scutellaria minor and the lovely little Campanula hederacea. In mosses and lichens these parts seem much less rich than the valley of the Fowey river.

In a garden about two miles from Carclew, on the shore of one of the branches of Falmouth harbour, I saw a myrtle fully 15 feet high, in the open ground ; it is of the small-leaved variety.

1841   *August* 16*th*.   Carclew.   Among the treasures of
Sir Charles Lemon's gardens I should have mentioned
Pinus (or Cunninghamia) lanceolata, and Arbutus
Andrachne, which grow beautifully.   Halesia tetrap-
tera is here a tree of considerable size; the
Hydrangea almost as fine as at Madeira, and as
often with blue as with pink flowers.   The variety of
the Pinaster which has been called Lemoniana from
having been first raised here, is very peculiar in
having terminal cones, or in other words a cone in
place of a leading shoot, in the manner of the
Proteas.   The flower garden looks nearly east, but
is sheltered by a noble grove of exceedingly tall trees
(chiefly firs) which overshadow a large tank.   I have
met here several remarkable persons : in particular
Colonel Sykes, the Indian naturalist, whose conversation is particularly interesting and instructive ; also
Colonel Yorke* (a very agreeable man), Professor
Ritter the geographer, Professor Lloyd of Dublin
University, *cum multis aliis*.   Two Miss Tremaines,
nieces of Sir Charles, are staying here ; the elder is
a fine, tall, handsome, showy girl, clever, and with a
striking boldness and originality of character ;
altogether a person not to be forgotten.   Miss Taylor,
whom I have also met here, is a charming girl.

August 17*th*.   I went into Falmouth with Sir
Charles, walked on to Pendennis Point, and spent a
considerable time in rambling among the rocks.
Pendennis Castle, celebrated in the great civil war,

---

* Brother of the wife of Sir Edmund Head.

has been modernized, and is garrisoned by one 1841 company of soldiers. It stands in a fine commanding situation, and is a conspicuous object from many parts of the neighbouring country, as well as from the sea. The shore is lined by rocks (of hard slate veined with quartz), not high, but strangely rugged, sharp, and jagged, and worn in some places into caverns and deep holes. The flakes of slate lie with their edges upwards, and these are so sharp that the rocks often appear almost like bundles of knives. I found here the usual plants of such situations, but nothing that was new to me.

Falmouth has a very picturesque appearance from the hill above Flushing, on the opposite side of the harbour; but it is a remarkably ill-built and dirty town, deficient in good shops and handsome buildings, and altogether disagreeable when one is in it. It is strange that the principal street of so important a town should be so narrow that it is with great difficulty two carriages can pass each other.

August 18*th*. From Carclew to Helston, a small town, standing on very uneven ground, like most of the Cornish towns, but better built than many of them.

August 19*th*. I hired an open carriage at Helston, and made an excursion to Kynance Cove and the Lizard Point. Kynance Cove is a place well worth seeing: the huge black frowning masses of rock, in some places standing out like giant castles, in others hollowed out into arches and caverns by the waves;

1841 the transparent beryl-green of the sea that rolls in
upon them, fringed with snow-white foam; the
smooth space of silvery sand between the crags;
the sea-gulls hovering about with their plaintive cry ;
make up altogether as fine a piece of wild coast
scenery as I have seen anywhere. There is something strikingly bold and massive in the forms of the
serpentine rocks. At a little distance they appear
intensely black, but when seen near, their richly
variegated tints of dark and light green, bright red,
black and white, add to the strange and beautiful
effect of the whole. Some of the caverns excavated
in them by the force of the waves appear truly fitted
to be the retreats of the sea-nymphs. Or, one might
fancy Kynance Cove to be the very place for Don
Juan and Haidee. The Lizard is a fine, savage,
rocky cape, but less peculiar and less interesting, I
think, than this cove, and I did not see it to advantage, as by the time I reached it a thick sea fog had
come on, and hindered me from seeing far. That
singularly local plant the Erica vagans grows in profusion all over the open moory land between Helston
and the Lizard, and is very ornamental. It seems
to be confined to this serpentine district, but is there
as plentiful as the common heath in any other part
of England ; yet in a garden it thrives perfectly well
without any serpentine. Immediately above
Kynance Cove it is abundant, but in a very dwarfish
state, not above three inches high. That spot is surprisingly rich in curious plants; I gathered, besides
this heath, Sanguisorba officinalis, Geranium

sanguineum, Arenaria verna, Scilla autumnalis, 1841
Genister pilosa (not in flower), Daucus maritimus,
and Erythrœa pulchella. The turf near the Lizard
light-houses is quite enamelled with the pretty little
Scilla, and there also I found Herniaria glabra.
Altogether this was a very satisfactory day.

August 20*th*. To Penzance ; the country open,
and of little interest till we come in sight of St.
Michael's Mount, which is a very striking object.
Marazion is a remarkably dirty town. The road
from thence to Penzance, three miles, runs along the
sandy shore, at a very little distance from high-water
mark, and northward of it extends a tract of country
lower than any other part of Cornwall, so that it is said
it would be easy to carry a ship-canal quite across the
county in this part. Penzance is one of the most
considerable towns in Cornwall, and has a picturesque appearance when seen from a little distance, but
can scarcely be called a neat or handsome town.
In the churchyard I noticed a tablet to the memory
of a man of the name of *Fudge*. In the afternoon I
walked along the shore to St. Michael's Mount, and
gathered by the way the Cynodon Dactylon, which
has been noted as an inhabitant of this locality ever
since Ray's time, and which I had previously met
with in Italy, at Madeira, Rio de Janeiro, Buenos
Ayres, and the Cape of Good Hope. It grows here
in the loose sand very near the sea, but not in great
plenty. Of St. Michael's Mount, which is a fine
object from all parts of the bay, I had formed a very
good idea from pictures and prints, and found it just

1841 what I had expected. The granite of which the island consists has the tendency, which seems usual in the Cornish and Dartmoor granite, to divide into massy cubical blocks, which stand out very boldly from its steep sides, and in some places overhang in a remarkable manner. Between these granite blocks, the ground is clothed with green turf, moss, and fern, and swarms with rabbits. The castle and church crowning the top of the rock make it a strikingly picturesque object. The view from it is extensive, and the wide sweep of the bay is pleasing to the eye, but the country about it wants variety and decided character; it is neither a rich country, nor yet a bold and rugged country, and presents in a general view nothing to attract the eye or excite the imagination.

August 21st. I made an excursion to the Logan Rock and the Land's End,—the latter about ten miles from Penzance. The coast scenery about the Logan Rock is exceedingly wild and grand: enormous blocks of granite piled one on another, like the ruins of castles raised by the giants, rise to a vast height, in many places overhanging the sea, which breaks on the foot of the cliffs. Many of the forms assumed by the granite rocks are very singular, and quite like artificial works on a colossal scale; some of them give one at a little distance the idea of Egyptian statues. The Logan Rock itself (which I have no doubt originally owed its peculiar form and position to natural causes) is perched on the top of one of the hugest of these granite piles, and is not easy to reach ; with the assistance of the guides I climbed 1841 up to it, the hardest scramble I have had for a long time ; yet many ladies have accomplished this feat. The stone has not been perfectly replaced in its original position, and does not rock as easily as it used to do. In itself it is a curiosity of which one may easily form an idea from description, and hardly worth the trouble of a visit, but the situation is sublime. The people say there was anciently a castle here (the place is called Castle Treryn), and they show some traces which may or may not be the remains of entrenchments. " Quien sabe ?" The scenery of the Land's End is of the same character and extremely striking. There is no beach or sand ; the Atlantic waves wash the base of the cliffs, which rise almost perpendicularly, naked and rifted, and worn into deep caverns by the fury of the sea ; and higher up, above the actual precipice, the great castellated piles of granite, hoary with lichens, jut out from the green brow of the hill. The Land's End itself is not so high as several of the neighbour-ing points, yet sufficiently fearful to look down from. At the time of my visit the weather was extremely fine, the sea almost calm, and bright blue, merely fringed with white foam where it broke upon the rocks ; many small vessels were in sight, with all their sails spread ; a few sea gulls on the wing, and numerous cormorants sitting and sunning themselves on the rocky islets near the cape. In stormy weather this must be a tremendous spot. The granite of the Land's End is porphyritic, with very large crystals

1841 of whitish felspar. The regularity with which it is divided, by horizontal and vertical fissures, into nearly cubical blocks of great size, as if it were stratified in two directions at once, is very remarkable; is this peculiar to the Cornish granite, or does it characterize the same rock in other countries? Certainly nothing of the kind is to be seen in the granite of the Cape of Good Hope, nor in the *gneiss-granite* of Brazil. These blocks, irregularly rounded and worn away at the edges and angles by the action of the weather, give its characteristic and striking peculiarity to the scenery of this part of the coast. No close investigation is required to distinguish the granite rocks from those of slate, or the serpentine from either, their general forms and colours indicate them sufficiently at a distance.

August 23rd. Return to Carclew,

August 27th. Visit the United Mines, one of the greatest mining establishments in Cornwall, about five miles from Carclew. The machinery is on a grand scale; the principal steam-engine is enormous, and is considered the best in the world, as doing the most work with the least proportional expenditure of fuel; it was made at Perran foundery, close to Carclew. The works of the mine extend upwards of a mile from East to West, and the refuse heaps are absolute hills. The ore is common massive yellow copper, no other being at present found in this mine; the operation of breaking it small with hammers, and of sorting it, is performed entirely by unmarried women and girls, whom one is surprised to see performing voluntarily such hard labour; they generally enjoy 1841 very good health, and become (it may well be supposed) excessively strong in the arms. The ore, when sold, is carried on a tram-road to a port on the North coast, and there shipped for Swansea. Several hundred women are constantly employed in breaking and sorting it.

The tract of country in which this and many other copper mines are situated (in the parish of Gwennap), is of a very peculiar and very desolate character : bleak, bare, stony moors, without a tree, or any vegetation except miserably thin and stunted heath ; the surface hilly, without any boldness of form ; small cottages of grey stone dotted here and there, with a few stone enclosures about them ; the absence of verdure as complete as in a Cape landscape ; tall steam chimneys and huge heaps of stone on every side, striking the eye at a distance. Yet this dreary-looking region is inhabited by a numerous, healthy, contented, and intelligent (though superstitious) population.

A party at dinner : Owen ; Professor Lloyd ; Mr. Sterling, a clever, animated and agreeable man, but devoted to the transcendental and mystical philosophy of the Germans, and consequently often obscure to ordinary comprehensions ; Mr. Charles Fox of Falmouth, a man deeply learned in various sciences, with his very pleasing wife. Owen's talk highly interesting.

NOTE.—Extract from the " Life of John Sterling,"

1841 by Thomas Carlyle:—" August 29th. I returned yesterday from Carclew, Sir C. Lemon's fine place, about 5 miles off; where I had been staying a couple of days with apparently the heartiest welcome. Sir Charles is a widower (his wife was sister to Lord Ilchester) without children; but had a niece staying with him, and his sister Lady Dunstanville, a pleasant and very civil woman. There were also Mr. Bunbury, eldest son of Sir Henry Bunbury, a man of much cultivation and strong talents; Mr. Fox Talbot, son I think of another Ilchester lady; and Professors Lloyd and Owen, the former of Dublin, son of the late Provost—a great mathematician and optician, and a discoverer in those matters. Owen is a first-rate comparative anatomist, they say the greatest since Cuvier, lives in London and lectures there. On the whole, he interested me more than any of them, by a force and downrightness of mind, combined with simplicity and frankness."

I have often heard my husband speak of the interesting conversations he had with John Sterling at Carclew.—F. J. B.

August 28th. Visited the iron foundery at Perran, in company with Sir Charles and the Foxes. The casting of a cylinder for a steam engine, a very curious and striking sight: intense red glow and heaving motion of the melted iron; jets of inflamed gas escaping from it; strange effects of light on the dusky faces of the workmen standing round.

August 29th. There are two Luccombe oaks in

Sir Charles Lemon's park, about 70 years old, and 1841 very nearly 90 feet high.

Mr. Talbot (of Laycock Abbey), who has been staying here for some days past, is a clever and very scientific man.

August 30th. The newspapers which arrived to-day bring the news that Ministers have been defeated, on the Address, by a majority of 91, in a very full house.

August 31st. Walked to Truro, six and a half miles, and back. It is the handsomest and best built town I have seen in Cornwall, and a striking exception to the general rule of narrow streets. Population, I am told, about 8,000. At the entrance of the town from Carclew is a pillar in memory of Richard and John Landor*, who were natives of this town.

September 5th. Yesterday evening at seven I left Carclew, where I had been entertained with the kindest hospitality, and came on by the mail to Launceston. This town has an exceedingly pictures-que appearance when viewed from the north, from the opposite side of the valley; its grey slate-roofed houses clustering up the steep side of a hill, the church tower rising above them, and the ruined castle crowning a high conical green knoll and overlooking the whole. This castle is supposed to be one of the oldest in Cornwall, but its history does not seem to be precisely known. The church, which is said to have been built in 1540, " at the sole expense of Sir Henry Trecarrell," is of

*The African travellers.

1841 granite, and every stone of it curiously carved on the outside. The streets of the town are rather narrow and dirty, and that which descends from the church into the valley remarkably steep. On the North side of the valley also the ground rises steeply, and is crowned by another conspicuous church—that of St. Stephen. The valley itself, through which runs a small rapid river, a tributary of the Tamar, is green and fertile, with a good deal of copsewood; to the East, the Dartmoor hills have a good effect in the landscape.

It appears that the Whig Ministry resigned (at last!) on Monday, and the Tories were regularly sworn in on Friday the 3rd : Peel, First Lord of the Treasury, Lyndhurst, Chancellor, Aberdeen, Foreign Secretary; Stanley, Colonial ; and Graham, Home Secretary; Lord De Grey Lord Lieutenant, and Lord Eliot Secretary for Ireland ; Goulburn, Chancellor of the Exchequer. So now we shall see what the Tories will do after all. I should have mentioned that the tower of Launceston church is not of the same date as the rest, but older, and not ornamented like the body of the church.

September 6th. To Oakhampton, a dirty little town, situated on the very skirt of Dartmoor, the first hills of that group rising immediately behind the town, on the South. The castle stands on a wooded knoll between the high road and the river, a little way out of town ; seems to have been large, but is quite in ruins ; the central tower, rising above the trees, on the top of the mount, looks well ; the rest nothing but fragments of walls almost hidden by 1841 verdure. The main tower square, not round like Restormel and Launceston.

I ascended one of the *Tors* of Dartmoor, the nearest to the town (Row Tor, I think, by the map,) and had an extensive view. The comparatively plain, enclosed and cultivated country on the North and East was strongly contrasted with the huge, bleak, moory hills stretching away to the South. There is something very remarkable in the aspect of the castellated granite rocks which crown most of these hills ; they resemble those at Trereen and the Land's End, though on a much smaller scale, and give a singular and rather picturesque character to the hills, otherwise lumpish and dreary. The horizontal and vertical divisions of the granite are as distinct and almost as regular as in the sandstone of the Cape of Good Hope. Not unfrequently the upper slabs overhang the lower, so as to suggest the idea of a hat, or the cap of a mushroom. I gathered many lichens on these rocks, but no mosses that were at all new to me. Trichostomum lanuginosum and Neckera curtipendula grow there in vast profusion, forming broad and thick cushions ; the former species with plenty of thecæ, the latter barren.

September 7th. Rain ; of which I will take advantage to note down a few particulars before omitted.

The western part of Cornwall is, on the whole, pretty generally enclosed and cultivated, yet the face of the country looks open and bare, for the fences

1841 are merely low banks of earth faced with stone, with nothing but a few furze-bushes or brambles on the top ; and there are scarcely any trees except about the gentlemen's houses. There is also a good deal of moorland, especially about the mines. Though a very large quantity of rain falls in the course of the year, the soil does not retain moisture, and the ground dries in a surprisingly short time after the heaviest showers, so that it is hardly ever muddy. The popular saying is that at Penzance it rains every day in the year but one, and then it freezes ; last winter, however, snow lay on the ground for more than a week. Sir Charles Lemon told me that at Carclew, even in the depth of that winter, the thermometer never was lower than 20 degrees Fahr.

Almost all the Cornish churches have square towers, which are very often handsome; a spire is rarely to be seen in the county. In the afternoon, go on by coach to Barnstaple, through Torrington and Bideford.

September 8th. Before starting, I walked about Barnstaple, and found it a good, handsome town, with a fine bridge across the river. Posted from thence to Ilfracombe, a delightful drive of eleven miles ; most part of the way we ascend a beautiful green wooded valley, the woods now beginning to assume the rich tints of autumn ; then cross some high ground, and winding down along the sides of very steep hills, look down on Ilfracombe. The combes or valleys which open to the sea in these parts are surprisingly deep and narrow, and the ridges between 1841 them terminate grandly in high abrupt headlands, presenting formidable cliffs to the sea. The shore is lined with rocks, singularly sharp and jagged, running out in reefs which may be compared to huge saws, or to magnified sharks' teeth, and which forbid all approach except in a very few places. It is a coast highly picturesque, but which one would much rather not approach by sea. A cove called (I think) Widmouth, a few miles East of Ilfracombe, is strikingly wild and romantic; yet I still decidedly give the preference to Kynance over everything else that I have seen on this coast.

The lower part of Ilfracombe, near the harbour, is dirty and ill-built, with narrow streets, but the upper and newer part is smart and pretty enough. It is overlooked by a steep green hill immediately on the North, at the foot of which is a little cove among the rocks, where people bathe; the lighthouse stands on a rock a little to the East of this, at the end of the neck of land which bounds the harbour.

The rocks here are of a grey shining slate, inter-stratified with a very hard thick-bedded rock which seems to be a compact kind of sandstone, and plentifully veined with quartz; the beds highly inclined. It is owing, I suppose, to the varying hardness of the different layers, and the consequently unequal degrees in which they yield to the battering of the sea, that these rocks present such formidable sharp points and edges—far more than any I saw on the Cornish coast.

1841 *September 9th.* More rain ; glorious weather for those who have an interest in importing foreign corn ! This was an odd sort of day—raining and blowing, yet very warm ; the air feeling like a steam-bath. I walked a good deal, in spite of the wet, but unprofitably. This neighbourhood certainly does not resemble Cornwall in the particular of the ground drying fast after rain.

I found Asplenium marinum in some plenty near Ilfracombe, but searched in vain for Adiantum Capillus, which I have heard has been found there.

September 10th. From Ilfracombe to Linton, twenty miles, the road being very circuitous ; the country uninteresting, till we descend into the valley of the West Lyn. The opening of this valley to the sea is more beautiful than I remembered. Having arrived at Linton, I lost no time in taking the well-remembered way to the Valley of Rocks, and with extreme pleasure saw again those magni-ficent crags with which I was so familiar ten years ago. All the associations of that time thronged back into my memory ; all the members of our party, our studies, amusements, disputes, rambles,— the very words and phrases that were used,—seemed as present to me as if it had been but yesterday. I thought much of poor Burgoyne, who (next to Edward) was my favourite associate among the party, and the frequent companion of my walks ; I thought of his placid good nature, of his unassuming manners and his easy inoffensive disposition, and I felt more regret for his death, I think, than I ever did before.

Strange is the force of local association even in a 1841 character so unimaginative as mine ! Yet the beauty of the scene and the brightness of the weather did not allow me to be melancholy, and only enough of regret mingled with my recollections to heighten my interest in the spot.

September 11th. Linton.—It is ten years since I was here before, and I have seen many countries and much beautiful scenery in that time ; yet this place charms me as much as ever. After all, the view as you look up from the sea-shore, taking in Lynmouth with the richly-wooded steep above it, ascending to Linton,—the entrances of the two valleys, and the high bold hills which bound them,—is of extreme beauty. The valley of the East Lyn perhaps did not strike me so much as at first, because I have since seen many like it ; yet it is delightful. The fine woods which clothe the upper part of it are now just beginning to be tinged with the brown of autumn. The lower part of the valley derives its chief beauty from the contrast of the two sides, the hills on the right bank being covered with oak copse, whereas those on the left bear only a few scattered thorn trees, are variegated with the different tints of fern, moss, and herbage, and in many places rough with rocks. This day was very warm and still, but during the greater part of it the sky was cloudy and overcast ; towards evening the sun came out bright. The Valley of Rocks appears to the greatest advantage at such a time, when the long shadows are cast from those grey fantastic crags, and the smooth

1841 turf of the lonely valley is lighted up with a rich golden glow.

September 12th.   A gloriously fine bright day, with a fresh breeze.   I strolled about the Valley of Rocks in the morning, and in the afternoon went on to Minehead, a twenty mile stage, which took three hours and a half, owing to the extreme steepness of the hills.   A formidable ascent from Lynmouth brings us upon high bleak moors (part of Exmoor), along which the road runs nearly on a level for many miles; then we descend another tremendous hill to Porlock, from whence to Minehead the way is through a rich cultivated plain, but still near the sea.

The prevailing rock at Linton is that hard, dark-coloured, stratified, somewhat granular and imperfectly slaty rock which I have mentioned as being, at Ilfracombe, interstratified with the regular slate. Here the strata are in appearance much less inclined than at Ilfracombe; quartz veins are less abundant, and there is much less of the genuine thinly-laminated slate; hence the diversity in the forms and outward appearance of the rocks at the two places. Though of so different a material, the crags of the Valley of Rocks remind one a good deal of the granite Tors of Dartmoor; but the slabs which compose them are thinner, more shivery and irregular, and their angles more oblique.   Nothing can be more wild or strange than the forms of these rocks, which, ranged along the high and narrow ridge between the valley and the sea, look like the ruins of some giant city of the earliest days. I wonder there is not some 1841 superstitious legend or tradition connected with so singular a spot.

September 13*th.* By coach from Minehead to Bridgewater, in three and a half hours ; thence by the railway to Bath. Day exceedingly hot.

September 14*th.* A heavy thunderstorm in the morning.

September 21*st.* Went from Bath to Clifton, and walked about the Downs and St. Vincent's Rocks ; much pleased with the scenery, which however I did not see to advantage, as there was a thick hot haze over the landscape. The weather has now (with the exception of the thunderstorm a week ago) been very fine ever since the 10th.

September 22*nd.* The gorge through which the Avon flows below Bristol is really very beautiful. The variegated red and grey tints of the limestone rocks are finely contrasted with the deep green of the rich woods which clothe the declivities on either side, and, together with the graceful curves of the Avon, and the green meadows and the broad Severn seen beyond, form most interesting combinations. The muddiness of the river is the only drawback. What effect the suspension bridge may have in the landscape when completed, I cannot tell, but the piers, which are all at present existing of it, are grievously ugly.

Asplenium Ruta muraria is very common in the crevices of the limestone rocks ; A. Ceterach much less so ; I observed also Rubia peregrina, Potentilla verna, and several curious Lichens. Heaths do not

1841 usually (in England at least) love a calcareous soil, yet Erica cinerea grows here in tolerable plenty on the Downs.

September 23rd.   Return to London.

September 25th.   To Barton.

October 19th.   I have noticed lately a singular sport or variation in the flowers of the common Arbutus, the corolla becoming polypetalous, sometimes with five petals, from the mere separation of the petals which in the regular state are united into an urceolate corolla ; sometimes an incomplete inner row of petals is formed by the partial conversion of the stamens, so that the flower becomes semi-double, and in this case the fertile stamens are proportionally diminished in number ; sometimes, on the contrary, the separation of the petals is imperfect, so that the flower appears to consist of two, three or four unequal pieces ; and a few flowers of the ordinary and regular structure grow on the very same bunches with these anomalous ones.   All these variations I observe on a bush in our arboretum,—almost the only Arbutus here which was not killed to the ground by the frosts of last winter.

Several of the Irish Yews here have borne fruit this year, the first time I have observed them to do so ; they are from 8 to 12 feet high ; the berries appear to me exactly like those of the common yew. Heath, our gardener, tells me that he has often sown the seeds of this kind, but that they do not come up true, continually sporting back to the original type of the common yew,— which is just what I should expect.

October 20th. The newspapers contain accounts 1841 of an extraordinary rise of the Thames at and below London, on Sunday last, the 17th. A great deal of mischief was done on the banks, and some persons narrowly escaped with their lives. The Temple Gardens were completely overflowed, so that a boat could pass over them; Palace Yard was under water, and the floor of Westminster Hall itself over-flowed. In this neighbourhood, in spite of the long continuance of wet and stormy weather, the waters do not appear to be unusually high.

October 26th. I finished the first volume of Elphinstone's " History of India", which has inter-ested and pleased me much. The principal part of this volume is occupied by a very full and satisfactory account of the religion, laws, customs, character, &c., of the Hindoos: in the first book he shows what was their state at the time of Menu's code (which he believes to have been compiled about the ninth century before our era) ; in the two next he describes the changes that have since taken place in their institutions and manners, their arts and sciences, and their present state. On the whole, he gives a view of the Hindoo character considerably more favourable than Mill's, and one which I am on every account more inclined to trust. Mill's book, able and philosophical as it is, leaves an unfavour-able impression of the author: it is the work of a most acute reasoner and profound thinker, but of one who delights to look at the worst side of human nature, and to put the worst interpretation upon

1841 human actions ; who is incapable of admiration
and enthusiasm, and prone to judge and condemn
all men by the rules of a moral system of his own,
without making allowance for the circumstances
in which they were placed. Elphinstone's, on
the other hand, is the work of a candid and
benevolent as well as enlightened man ; and besides,
his long and intimate acquaintance with the natives
of India gives great weight to his opinion of them,
especially when that judgment is favourable. He
shews very clearly why we ought to receive with
caution the opinions formed of the national character
by the generality of Englishmen in India, in particular by missionaries, judges, police magistrates,
and officers of revenue. At the same time, he seems
to be quite free from any unreasonable partiality, or
disposition to exaggerate the good qualities or the
civilization of the Hindoos.

Elphinstone's style, though not of extraordinary
merit, is far more easy and agreeable than Mill's, and
some of his descriptions are very picturesque ; in
particular, that of the march and encampment of a
Hindoo army (book II, chap. 2), and that of the festival
of Rama, as celebrated by the Marattas (b. III,
c. 11).

One of the most striking things, to me, is the
great change that Elphinstone shows to have taken
place, since the time of Menu's code, in some of those
very points of the Hindoo manners and institutions
which we have been used to consider most invari-
able. In the first place, the religion taught by Menu,

as well as by the still more ancient Vedas, is decid- 1841
dedly *monotheism* ; the doctrine of the unity of God is
repeatedly inculcated : and though "the three principal manifestations of the Divinity (Brahma,
Vishnu, and Siva), with other personified attributes
and energies, are indeed mentioned, or at least indi-
cated, in the Vedas," yet they are but rarely and
slightly mentioned, and not as objects of any special
worship ; no allusion is made to any of their incarna-
tions; there seem to have ·been no idols, and no
worship of deified heroes. All the extravagant and
monstrous fables of Hindoo mythology, of which we
have so often heard, are first mentioned in the
Puránas, which Elphinstone thinks were composed
between the 8th and 16th centuries of our era,
"although in many places from materials of much
more ancient date." Again, the laws of Menu allow
to a certain degree, the intermarriage of different
castes, which is now strictly, prohibited ; they permit,
and on certain occasions even enjoin, the eating of
most kinds of flesh, nay, even of beef ; they make
no allusion to the self-immolation of widows, but
rather imply that it was a practice unknown in those
early times. The numerous sects of religious mendi-
cants (Gosayens), likewise, have all arisen since the
days of Menu. On the whole, Mr. Elphinstone is decid-
dedly of opinion that the Hindoos were once in a
higher condition, both morally and intellectually than
they are now.

The excessively scrupulous respect for animal life,
which has often been attributed to the Brahmins,

1842 appears really to belong to the priests of Buddha (Elphinstone, b. II., chap. 4).

———

From the 26th of October, 1841, when he was at Barton, till the 18th of June, 1842, when he writes from North Wales, I can find neither journal nor letters. I believe he spent the winter at Barton, and part of the spring in London.—F. J. BUNBURY.

———

June 18th. To Beddgelert and Llanberris, 24 miles.

The day was cloudy and dull, so that the beautiful scenery betweeen Tan y Bwlch and Pont Aberglaslyn did not appear to so much advantage as I remember it did in Sept., 1832, when I first saw it. A succession of noble Alpine views all the way from Beddgelert (or indeed from Pont Aberglaslyn) to Llanberris, especially after entering the pass of the latter name, which is truly sublime. Such a scene of black crags and frowning precipices, serrated ridges, and bare shattered rocks, as I have scarcely looked upon since I was in the Alps; it beats even the Gap of Dunloe, which is savage enough. The mountains rise all but perpendicularly on either side of the torrent, almost bare of vegetation, and their tops are most singularly jagged and toothed. The summit of Snowdon itself is not seen from the pass, though we have a fine view of it from a part of the road some 1842 miles nearer to Beddgelert; but the very sharp, abrupt, naked peak called Crib coch, which frowns directly over the pass, is, I believe, considered as a part of Snowdon. There is an excellent hotel at Llanberris, situated nearly between the two lakes, at a very short distance from the old ruined tower (it scarcely deserves to be called a castle) which over-looks the upper lake. It is a fine Alpine view, looking up this dark little lake, between the steep and savage mountains which hem it closely in. The shores of the lower lake are of a much tamer character, the mountains rapidly diminishing in height, and opening out wider, as you descend the valley. There are very extensive slate quarries near Llanberris, on both sides of the upper lake.

June 19th. Llanberris.—I strolled up the pass and spent some time in botanizing, but found nothing remarkable except Oxyria reniformis, Polytrichum Hercynicum, and one very poor specimen of Saxifraga stellaris.

June 20th. I ascended Snowdon, with an ex-perienced guide, and had a very satisfactory excursion. The ascent is very much easier than I had expected· for the greater part of the way it is a horse-path, rough and stony but by no means steep, winding along the green, sloping sides of huge hills, —the reverse sides of those very hills which towards the N.E. frown over the lake and pass of Llanberris in such bold cliffs. There is very little heath on these mountains: what is not bare rock is covered

1842 with fine short close grass (chiefly Festuca ovina)
mixed with moss. The first part of Snowdon proper
that we approached was the naked black precipice
called Clogwen du r' Ardu (I am not sure of the
spelling), at the foot of which is a very small lake of
singularly deep blue water, and near this a copper
mine. For some time we proceeded directly towards
this cliff, then turning, and leaving it on the right,
had a pretty steep pull for some way up, but nothing
like scrambling ; this brought us to a stone enclosure,
where we left my pony. From hence it is about
half a mile to the summit, the path running along
a stony ridge, not very narrow, with a steep but not
precipitous descent on the right hand, and on the
left a sheer precipice of vast depth, from the sides
of which the shivery, sharp and shivered rocks
stand out like knife-blades.

When I reached the top of the mountain (which is
crowned with a pyramid of stones), I could not see
ten yards in any direction : we were enveloped in
impenetrably thick white clouds, which came rolling
up before the wind in surge-like volumes, forcibly
reminding me of the scene in " Manfred." Thus we
remained for a considerable time, the mist now and
then opening sufficiently to give glimpses of precipices and lakes ; till at length, just as we were beginning to descend, it fairly cleared away for awhile, and
I had a fine view. The distant part of the landscape
seen from hence is, as in all similar cases, unsatisfactory, but the near view, of the Snowdon group
itself, is very grand. Directly beneath me, at the

bottom of a crater-like abyss of fearful depth, lay 1842 the dark pool called Llyn Glas, the source of the river which we cross at Pont Aberglaslyn : it is nearly surrounded by naked and rifted precipices, rising on the one hand to the main peak of the mountain, on the other to the lofty, sharp and serrated ridge which runs to the Llanberris Pass and ends in the remarkably sharp peak called Crib côch (or Grib gôch, I am not sure which). Of the four principal branches or ridges which run out from the central mass of Snowdon, this is the highest and most remarkable : the most elevated point of it, which rises into a sort of hump, is called, as I understood my guide, Girn lâs, and the sharp ridge, serrated like a shark's jaw, which runs from this to the Crib côch bears the name of Crib Ddescil (pronounced Deskil). The reddish and, as it were, scorched hue of the steep and naked sides of this ridge, contributes to give to the vast hollow, at the bottom of˙ which lies the Llyn Glas, the aspect of a volcanic crater. The other principal branches of the mountain are, South-ward of Crib Ddescil, the very sharp ridge called Llywart (as well as I could make out the name) ; more to the West, extending towards the sea, the Clawdd côch, or 'red wall'—not ill-named ; and lastly, in the direction of Carnarvon, the ridge which on its North face forms the black precipice that I mentioned in ascending, but on the other side sinks in a smooth green slope. Between all these are numerous small lakes, the number of them that lie in the hollows of Snowdon being altogether very considerable.

1842   I procured here several good plants.   The beautiful Silene acaulis, with its bright rosy blossoms now in full perfection, grows in several places among the rocks very near to the summit; Saxifraga stellaris is abandunt in the same situations; Saxifraga nivalis, a much more rare but less pretty species, occurs here and there in the fissures of the almost inaccessible crags.   This last I did not myself gather, the places in which it grew being out of my reach, but the guide got it for me, as well as Saxifraga hypnoides, Thalictrum alpinum Rhodiola rosea, and Salix herbacea.   Carex rigida (I believe) is plentiful on and near the summit, but I could find no good specimens of it.   Altogether, I found this excursion very pleasant, and not at all fatiguing.   Snowdon is the sixth mountain I have climbed, the others being Cader Idris, Mangerton, the Brevent, the Corcovado, and Table Mountain.   The last is very nearly of the same height as Snowdon, but far more difficult to ascend, and the view from it is nothing like so fine.

June 21*st*.   At Llanberis.   A very changeable day, with much rain in the morning and evening.

June 22*nd*.   From Llanberis by Caernarvon to Bangor. The view from the road along the shore of the lower lake, looking back towards the Pass, was very fine, as the clouds threw broad, dark masses of shadow over the rugged mountains, and the old tower, standing on a sort of promontory pushed forward between the lakes, was a picturesque object in the middle distance.   The topmost peak of Snowdon is well seen from this road, when you are a mile or two

from Llanberis, as well as the black cliffs of Clogwin 1842
du r' Arddu.

Hills of moderate height, but very rough and
rocky continue along both sides of the lower lake,
and a little way below it; then we enter on a nearly
level country, enclosed and cultivated, which continues to Caernarvon. The old castle here is a
noble pile of building, though not remarkable in its
situation. From hence to Bangor the road runs
between hedges, through a rich country, mostly in
view of the Menai Strait and the isle of Anglesea;
the latter not altogether flat, but tame and cultivated.
I had before had a good bird's-eye view of nearly the
whole island, from the top of Snowdon. Its surface
seems to be very little varied. The Menai Bridge
is a splendid work, one of the finest monuments of
modern science and enterprise, and really beautiful
as well as useful and vast.

June 23rd. Bangor can boast of an excellent inn,—
the Penryhn Arms,—but in other respects it does
not take my fancy: it is true that the view
across the strait to Beaumaris is pretty, and looking
along the coast to the northward, Penmaenmawr
and the Great Orme's Head are very picturesque
objects; but there seem to be no good walks in the
neighbourhood of the town. The mountains are too
far off, and the shore is nothing but rocks and mud,
fit neither for walking nor bathing.

June 26th. Went by coach to Capel Curig,* in

---

* Here he met Mr. Charles Darwin and spent a little while with him with
much enjoyment of his conversation, ₱ J. B.

1842 very cold, gloomy, blustering weather, which chilled
my enjoyment of the scenery.   Yet the rocky moun-
tain-pass leading to Llyn Ogwen is magnificent, the
mountains amazingly rugged and savage.   I was
struck with the vast extent of the slate quarries at the
entrance of the pass : the whole side of a mountain
is cut into terraces and escarpments.   After I reached
Capel Curig the weather improved for a time, and
the sun shone out, but the wind continued to blow
with surprising fury all day, so that the little lakes
were all in a foam, and broke upon their shores in
actual waves.   Snowdon, of which there is a capital
view from the garden of the inn, looked black and
sullen, but his outline very distinct, and not obscured
by the clouds.   Seen from hence, he appears to
have three sharp and well-marked peaks.

June 27th.   A much finer day.   I walked three or
four miles up the Llanberis road, to where it is
joined by that from Beddgelert, and then climbed a
high hill (a sort of offset of the Glyder), from which
I had a good view of three different valleys,—the
Pass of Llanberis, the Capel Curig valley, and the
much prettier and more fertile one (Nant Gwynant)
running down to Beddgelert.   The hill in question
exactly commands the junction of these three.   In a
depression or slight hollow between it and the
Glyder, lies a small, dark, sullen-looking pool, from
whence issues the stream (a rapid and beautifully
clear one) which runs down to Capel Curig and
through the lakes.   Snowdon appeared to great
advantage to-day, the strong and variable lights and

shades produced by the passing clouds, giving a 1842
fine effect to his bare rocky ribs and deep hollows.
From the hill aforesaid I had a noble view of Crib
côch, of the Wyddfa itself, with the gloomy precipice beneath it ; and of the long ridge, jagged like a
cock's comb, which runs out from it to the left, and
which is called, I believe, Lluwydd.

I found no very remarkable plants these two days ;
abundance of mosses and lichens, but none altogether new to me. Lycopodium Selago grows in very
great plenty on the hill I ascended ; L. selaginoides
is not uncommon in bogs near Cape Clurig. The
viviparous form of Festuca ovina very common.

Most parts of the surface of these mountains,
where not bare rock, are boggy, and in such
situations (not only here, but in North Wales
generally,) the prevailing plants are :—Juncus
squarrosus, Melica caerulea, nardus stricta,
Eriophorum angustifolium, Scirpus cæspitosus,
Carex stellulata and some others of that genus ; our
two species of Pedicularis, and those beautiful plants
the Narthecium, Pinguicula, and Erica Tetralix.
The most common mosses are Dicranum flexuosum
(especially the tall black variety of it, which is
always barren), Trichostomum lanuginosum, and
others of that genus, Bartramia (?) arcuata (barren),
and the Sphagna.

June 28th. Another odious day, blowing furiously,
with a thick mist and drizzling rain. I walked to
Llyn Ogwen by the old road, got very wet, and saw
little or nothing. Lobelia Dortmannia was the only

1842 remarkable plant I met with, and that hardly yet in perfection.

June 29th. Ascended Moel Shiabod, the long mountain which bounds the Capel Curig Valley on the South, and which according to the map is 2,878 feet high. The first part of the ascent is through a pleasant open wood of birch trees, which is succeeded by pastures interspersed with rocks; about the middle of the mountain there are troublesome bogs; higher up, a fine elastic mossy turf, very pleasant to walk on; and lastly, the actual ridge is covered with great angular fragments of rock in strange confusion. The reverse or southern face of the mountain is nearly precipitous, in many parts quite so, and one looks down into one of those deep crater-like basins which are frequent in the mountains of North Wales. The view from the highest point is extensive and, in the West and South West directions, very fine, embracing a multitude of vast mountains in many successive gradations of distance. Snowdon looks exceedingly majestic: he appears with two peaks of nearly equal elevation, namely the Wyddfa and Crib Ddescil; for Crib côch, which makes the third in the view from Capel Curig, is so much lower as to make no figure from this elevation. I had a good view of Llyn Llydau, the largest lake on Snowdon, and could discern Llyn glâs, the little lake lying in the crater beneath the Wyddfa. Many other small lakes are visible in different directions. I counted nearly a dozen altogether. Far off I thought I could distinguish Cader Idris, but of this I am not sure,

Lycopodium Selago extremely abundant on Moel Shiabod, from the middle region upwards ; L. Alpinum common near the top, among the short grass ; Empetrum nigrum in the middle and upper regions ; Saxifraga stellaris in the clefts of rocks at the top, and in great profusion in some little mossy rills which trickle down on the Capel Curig side.

On and near the summit I gathered many curious mosses. Altogether this was the pleasantest and most satisfactory day that I have passed since the 20th.

June 30th. Walked to Llyn Idwal, a wild dark lake lying in a deep hollow, almost surrounded by black craggy mountains. It is a fine savage scene, much like the *craters* of Cader Idris, with a larger lake. The lower parts of the mountains, though exceedingly steep, are partly clothed with moss and a little verdure ; above are magnificent precipices, strangely rifted and shattered. A deep cleft or chasm, singularly black, is conspicuous in the highest part of these cliffs ; it is called, I believe, Twll dû.

Vast quantities of angular fragments of rock lie on the steep slopes from the precipices down to the edge of the water, disposed as it were in streams, as if they had been poured down from the top ; Darwin takes them for the moraines of an ancient glacier.

The effect of the whole is gloomy and grand, especially when viewed from the head of the lake ; yet it is not so wonderful as it has sometimes been represented ; I think there are other scenes in North Wales of a higher degree of savage grandeur. The

1842 distance from Capel Curig is something less than six miles; one follows the high road as far as a little public house at the western end of Llyn Ogwen, and then turns off to the left, by a rough but not steep mountain-path, which brings one after half a mile or a little more, to Llyn Idwal, at the lowest part, or lip as it were, of its basin, where a stream issues from it and tumbles down the mountain into the Ogwen river.

I found one rare plant here, Asplenium viride, which has long been noted as an inhabitant of this spot; it grows in the clefts of the rocks about the head of the lake, but is not easy to get at. I had not seen it before since I was in the Tyrol in 1828.

I observe in North Wales, that very few ferns occur at any considerable elevation on the mountains: the only one I saw very high up was Pteris (Cryptogramma) crispa, near the summit of Snowdon. On Moel Shabod, above the middle region, I met with none but Blechnum boreale, and that very sparingly and stunted. The common Brake does not ascend to any considerable elevation. Of the flowering plants, Galium saxatile abounds from a low level to the very tops of the mountains. The Foxglove, the Cotyledon, and the very pretty Sedum Anglicum, are some of the most conspicuously common plants in North Wales.

July 1*st.* I took my leave of the mountains, with considerable regret, and went by mail-coach to Shrewsbury. Such a mode of travelling does not allow one to observe much. I could see, however,

that the country for a good many miles from Capel 1842
Curig, indeed the greater part of the way to Cernioge,
has a great deal of beauty : thence onwards it is
open and dreary till one enters the valley of Llan-
gollen. About Chirk, or a little beyond, it finally
loses the Welsh and assumes the English character.

Shrewsbury has something of an antique and peculiar aspect that is rather striking.

July 2nd. From Shrewsbury by coach to Birmingham, and thence by the railway to London. Nothing
particularly interesting in this journey except the
strange and horrid-looking country between Wolver-
hampton and Birmingham, a region of coal-pits, iron
mines, forges and furnaces, of fire and smoke and
blackness, the inhabitants looking as black and grim
as their country.

July 6th. Return to Barton.

July 20th. After two or three days of most oppressive stifling heat came a heavy thunder-storm,
with such torrents of rain as I have seldom seen in
this country: the Eastgate street at Bury was completely flooded, so that the water came in at the doors
of the houses and the Northgate street was converted
into a stream which rushed down like a mountain
torrent.

July 27th. I have lately read some important
letters received by my father while Under Secretary
of State, and which he has preserved : several of them
are from officers engaged in the battle of Waterloo,
and among these the best is from Sir Peregrine
Maitland, giving a very clear and spirited account of

1842 the operations of his brigade of Guards. Another is an interesting narrative of the battle of Salamanca, by Colonel Cathcart ; but that which had most novelty to me was the information collected by Sir Herbert Taylor concerning the battle of Jena, from the accounts of various Prussian officers. From these it clearly appears that, whatever might have been the blunders or the imprudence of the Prussian Generals, the rout was owing mainly to the very bad behaviour of the soldiers. The infantry of the King's corps (which was opposed to Davoust) stood still, and all that their generals could do could not persuade them to advance towards the enemy ; their cavalry fled at the first check, and quitted the field altogether.

August 4th. I drove over to Cambridge, and spent some hours in rambling about the college walks and Granchester meadows.

August 5th. Walked to Cherry Hinton and the Gogmagog hills, in order to botanize, and met with one very rare plant, for which this locality has been noted ever since Ray's time,—the Athamanta Libanotis. I found it growing in several places by the sides of the road from Cherry Hinton to the Gogmagogs, most plentifully in an old chalk pit overgrown with herbage, on the right of the road.

The only other remarkable plant I gathered was Galeopsis versicolor. I wanted to get the pods of Hippocrepis comosa, but was too late for them.

August 6th. A mild wet day. The rain did not prevent me from walking to Madingley, which with its park and quiet little churchyard and solemn yew

trees, is the only place in the neighbourhood of Cam- 1842
bridge (beyond the college walks) that can by any
means be called pretty. The college walks are certainly charming; especially those at the back of
King's and Clare Hall; and now in the vacation,
above all, there is a stillness about them peculiarly in
character. I like to lounge under those venerable
trees, and call back to memory the days when I wore
the blue gown of Trinity, and when a stroll in these
walks was not merely the amusement of idle hours,
but a relaxation after hard study. I do not think,
in reality, that I had more enjoyment of life in those
days than I have now, nor are the years that I spent
at Cambridge in the number of those which I would
wish to live over again (there are some which I
would, very gladly); but still I have a pleasure in
recalling them.

August 7th. Went in the evening to Ely.

August 8th. Spent the morning in viewing Ely
Cathedral, which is truly a noble and magnificent
building. Built at different times, it exhibits many
various styles of architecture, but the effect of the
whole is truly grand. The great Western tower, with
that which flanks it, is particularly fine; the
corresponding one (which should have flanked the
great-tower on the other side) is unfortunately wanting; the West door and porch, of a much later date
than these towers, are extremely beautiful. One of
the small chapels in the interior (Bishop West's
chapel, I think) is a singularly beautiful specimen of
the most florid Gothic; the elaborate richness and

1841 lace-like delicacy of the ornamental work are quite surprising. Most of the chapels, however, suffered more or less from the barbarous fanaticism of the Puritans, who destroyed or defaced most of the figures in the niches, and broke much of the carved stone-work. In most of our fine old cathedrals we have to lament similar ravages. Politically, I respect and admire the Puritans as the brave assertors of our most valuable rights and privileges; but I despise their fanaticism, and detest the barbarity with which they defaced or destroyed the most beautiful and venerable works of art. " There is a superstition in avoiding superstition," says Bacon; and this was never more remarkably exemplified than in the horror of the Puritans against "superstitious pictures"; images, paintings, and ornaments in churches, church music, and a variety of other things equally innocent. I returned from Ely to Barton by way of Mildenhall. Nothing in the world can be more drearily ugly than the country between this latter place and Ely.

August 10th. Mr. Thornhill tells me that a black stork (?) (Ciconia nigra), a bird extremely rare in England, was lately killed on the edge of Romney Marsh, and is in his or his father's possession. He also told me of a Bustard having been seen near Riddlesworth not more than a year or two ago. This bird was generally supposed to be quite extinct in England.

The newspapers contain a most interesting and beautiful letter from Sir Robert Sale, giving the whole account of his heroic defence of Jellalabad in 1842 the face of every disadvantage and danger. It was an exploit comparable only to Clive's defence of Arcot, and ranks Sale high among the many noble-spirited and heroic men who have adorned our Indian history.

August 18th. Weather amazingly hot and dry ever since my return from Cambridge, especially these last five days, which have been perfectly Italian and cloudless. We have had a wonderful summer altogether. Patrick Blake,* who is lately returned from China, where he was actively employed in the naval operations, says that the trade is carried on with more activity than before the war, the local authorities being either unable or disinclined to check it. He confirms my belief that the pretence of morality which was put forward by the Chinese government as the ground for interfering with the opium trade, was a mere humbug, and that the real object was to stop the exportation of silver. The Chinese have learnt to construct very formidable works, which would be all but impregnable if properly defended : at Amoy their new batteries were nine feet thick of solid blocks of granite, besides a great thickness of earth. But they have no skill in gunnery, and their guns, mostly of very large calibre, are mounted in such a way that they cannot traverse, and can fire only in one single direction. The guns also are very apt to burst, either from badness of quality or from overloading; at Amoy or Ching-hae (I

* His cousin Admiral Blake.—F·J.B.

1842 forget which), when the English entered one of the batteries, they found that five guns out of nine had burst, and done great mischief to their own people. At Ching-hae were found a number of guns newly cast, not of iron nor of brass, but of pure copper, which a very few rounds would have melted at the touch hole. The Chinese, though so deficient in the art of offence, will stand fire well enough behind their walls, but as soon as they see the red coats advancing, they immediately take to their heels, and have no idea of rallying after a defeat.

Patrick Blake says that the Chinese, though not a sanguinary or actively cruel people, are very callous and indifferent to human suffering: if one of their own people meets with an accident, breaks a leg or an arm for instance, they pass him by without any notice or the least thought of assisting him. Nothing surprised them more than the care which the English took of the wounded after an action.

August 20th. The accounts from the manufacturing districts continue to be very unpleasant ; the insurrection of the working men is very widely spread, and as fast as it is put down in one place it breaks out in another. Fortunately the insurgents do not seem to have any good leaders, nor have they anywhere shown much resolution, but it requires a vast force to defend so many points as are at once exposed to danger. It is only in Staffordshire that they have as yet done much real mischief to proerty, or shown a violent and malignant spirit. It is not easy to make out the real nature and origin of this

rising; its *apparent* origin was a most ill-judged 1842
reduction of wages by some few masters, which
occasioned a strike among their workmen; but the
almost simultaneous disturbances at so many distant
points lead one strongly to believe that there must
have been an organised and preconcerted plan of
resistance. The professed object at first was merely
to obtain better wages, and even now, in some places,
the " turn-outs " adhere to this declaration, openly
disavowing all political objects. But in general the
Chartists, if they did not at first organise the insurrection, have taken advantage of it, and are trying to
extort concessions by the terror of civil war. Those
villains Feargus O'Connor, Mac Douall, Cooper, and
others, have been active in spreading their poison
among the multitudes, and will I hope be punished
as they deserve. It is not wonderful that the working men of Lancashire and Yorkshire, who have so
long been suffering extreme distress and privation,
should be the ready tools of any artful demagogue;
still less can we wonder that the Staffordshire colliers
should be ready for any violence and outrage, when we
read in the reports of Lord Ashley's Commission, of the
kind of life they lead, and the treatment of their women
and children. The conduct of the government hither
to seems to have been judicious, and that of the
soldiers exemplary, and I trust the sedition, though
extensive, may ultimately be put down without any
very serious mischief. In some places the workmen
themselves have defended the mills, and repulsed
the rioters who wanted to force them to join the strike.

1842 *August* 21*st*. This hot dry summer has made the Bignonia radicans in our Arboretum blossom much more freely than usual, so as to be really very handsome, which in general it hardly is.

Magnolia grandiflora is also in remarkably fine bloom ; and the Ilex (for the first time with us) has abundance of young acorns. The wheat harvest in this neighbourhood is nearly finished, and is a very fine one.

August 25*th*. Patrick Blake tells me that during the seige of Chusan the great city of Ningpo (which was not then in our possession), lying nearly opposite to Chusan, and but a short distance from it, the Mandarins were continually sending over spies to note and report all such as held communication with " the barbarians ", and who were marked out for punishment on the restitution of the place. Ningpo, he tells me, has above 400,000 inhabitants.

I find Blake has a very high opinion of Lord Auckland, whom he reckons one of the best Governors-General that India has ever had, and he is disposed to look on the censures to which Lord Auckland is now subjected on account of the unlucky Cabool expedition, as merely " prophecies of the past ". He admits, however, that the Duke of Wellington and Mountstuart Elphinstone, two very high authorities, condemned it from the first. So, I know, did Colonel Sykes, at least long before the disasters occurred.

August 27*th*. Very bad news from the Cape. A small force, not more than two companies of the 27th,

with a few guns, had been sent to take possession of 1842
Port Natal ; the emigrant Boors refused to submit ;
the soldiers attacked their camp, and were repulsed
with loss, leaving behind them two guns. I am sorry
for this, in the first place because it will be a greivous
vexation to my dear good friend the Governor ; and
likewise because the honour of our arms is compro-
mised, and it will no doubt be thought necessary to
follow up the matter, and engage in a serious war
with these people, which will cost more bloodshed
than many people in England are aware of. I am
afraid the commander of the expedition, as is apt to
be the case with regular soldiers, undervalued his
opponents because they were not regulars.

The Cape Boors are formidable fellows : they
have a stubborn, sullen, dogged determination of
character ; they are not without a sort of discipline
and tactic of their own, are admirable marksmen,
and use very long guns, which carry much farther than
a soldier's musket. The country about Natal, too,
appears to be remarkably well suited to bush-fighting.
I fear we shall hear of further disasters, for the force
at the Cape is too small to spare such a number of
men and guns as would ensure success. It is true
that the Amaponda and other Caffer tribes would no
doubt be willing enough to join us against the Boors,
—but I am not sure that they would be of much use
as allies. I trust that Sir George had positive orders
from the Government at home to make this attempt
upon Natal, and that it was not a stroke of his own.
It seems to me a great error : I never could see the

1842 policy or the reason of meddling with these emigrants, who had quietly rid us of their presence, and had neither committed nor seemed likely to commit any act of hostility towards us.

September 1st.   Captain Franks, who is just come down from London, says it is generally believed in the City that the Income Tax will produce above seven millions, instead of four, as Peel calculated. A striking proof, if it be true, of the vast wealth and resources of this country.

September 4th.   We had a visit yesterday from Mrs. Buller,* who brought with her the famous Carlyle, a man I was very glad to make acquaintance with.   I took a great fancy to him.   He is a mnch younger man than I had imagined,—not forty, I should think; of middle height, thin and dark, with very fine eyes, and an expressive countenance ; his manners perfectly simple and unaffected, and without the least pretension.   He strongly impresses one with the idea of a thoroughly sincere, honest, and earnest man. We had much talk about the state of society, the manufacturing system, and the poor, and his remarks seemed to me exceedingly just and sound as well as benevolent.   I like his conversation much better than his books.   He talks as good, plain, clear intelligible English as one would wish to hear, instead of the strange obscure jargon of Germanized English which he writes.   His Scotch accent is very strong. Mrs. Buller told me that when Carlyle was writ-

---

* She was the Mother of Charles Buller and was staying with another son Reginald who had a living at Troston in the neighbourhood of Bury.

ing his history of the French Revolution, and had 1842
finished more than half of it, his manuscript, which
he had negligently left exposed, was destroyed, by a
stupid housemaid ; above a year's work destroyed at
once ; and he had to write the whole over again.
She says he bore this misfortune with the utmost
courage and philosophy.

September 8th. I lately visited Mr. Richards, and
saw in company with him, the interior of Stowlangtoft church, which is handsome, and contains very fine
carved woodwork of great antiquity : the backs of
the seats are all of old oak or chesnut, elaborately
carved, and in great variety. Some of the woodwork
which was defaced has been restored with much
taste and judgment under Mr. Rickard's* direction.

There is a curious monument of a Sir something
D'Ewes, the father of the noted Sir Symonds
D'Ewes ; kneeling figures of himself and his two
wives and all their children, painted of the natural
colours, like the monuments in Framlingham church.

Mr· Richarks is a man of primitive simplicity of
manners, very unworldly, very zealous in the pursuit
of knowledge, and much devoted both to antiquarian
researches and to natural science, though rather an
amateur than deeply scientific. His religious
opinions, I believe, are those of the Puseyite or new
Oxford school.* His wife is a remarkably pleasing
person, and a good botanist.

* The Rev. Samuel Richards, who has often been called the Keble of the
Eastern counties. F.J.B.

* I think he was a High Churchman but that he did not belong to the
Puseyite party, F.J.B.

1842   *September 20th.* The last accounts from the Cape
are more satisfactory. Captain Smith, with his
small force of the 27th, though closely besieged in
his camp by the insurgent Boors, and reduced to
great distress for provisions, held out gallantly for
upwards of a month, till relieved by the troops
sent from Cape Town under Colonel Cloete.
These, very fortunately, effected their landing
without any serious loss, though under a heavy
fire from the Boors, and drove the latter from
their positions. It seems to have been a very
well-managed affair. The Boors are reported
to have withdrawn from the neighbourhood of
the port, and marched to join the great camp of
their friends, higher up the country ; but the Caffer
tribes, seeing them engaged with us, took the
opportunity to renew hostilities, and had cut off
some of their stragglers. They, however, (the Boors I
mean) do not show any inclination to yield, and their
character is so stubborn and doggedly resolute, that
I have little doubt, if they find themselves unable to
make head against our forces, they will retire far
into the interior of the country, even beyond the
tropics, rather than submit to British authority.
But I hope our Government will not persist in keeping mere military possession of Natal. It is probable
that sufficient troops cannot be spared to occupy it
in force, and to leave a small detachment there would
only be exposing that detachment to be destroyed
by the Boers. It would be much better, now that

the honour of our arms has been vindicated, to with- 1842
draw our troops at once and give up the place altogether, unless our Government resolve to colonize it,
which would perhaps be the best plan of all.

October 8th. The lovely Belladonna Lily is now
blossoming magnificently in front of our conservatory,
having more flowers open at once in each umbel
than I ever observed in its wild state. Nothing can
exceed the delicate beauty of its colour.

October 23rd. I have been a week in London,
walking about the streets during the day, seeing all
I could see. The principal sights now to be seen
are the Chinese Exhibition and the Panorama of
Cabul, both very well worth visiting.

October 25th. To Leeds, by the railway, in ten
hours. On the way I read in the newspapers the
last accounts from the Cape. The Natal affair is
happily settled, the emigrant Boers made their submission to Colonel Cloete, gave up all their prisoners
and the guns they had taken, and acknowledged the
authority of the British government; in return for
which Cloete grants them an amnesty (with the
exception of four or five ringleaders,) allows them to
retain the lands they had occupied, and their own
administration and institutions, till further orders
from home, and promises them protection against
the Zulus. It seems that the shopkeepers of Cape
Town, and some of the Cape newspapers, are making a most furious outcry against Cloete, for his
moderation and leniency towards the Boers. This
is really very unreasonable; his conduct seems to

1842 me both wise and humane. It remains to be seen what our government will decide on, I think they have only two eligible measures to choose between : either to withdraw the troops, and leave the Boers entirely to themselves, (satisfied with the acknowledgment of our supremacy,) or to convert Natal into an integral part of the Cape colony.

I confess I had entirely mistaken the character of the Boers. I believed them to be a brave and resolute people ; and had no notion that they would so readily give up the struggle, and acknowledge the authority they had so lately defied. Clarke, who knows them right well, says that they are in fact cowards, and never willingly fight except at a great advantage. He thinks that Captain Smith, from ignorance of their character, rashly incurred a loss which might easily have been avoided ; indeed, that if he had understood the people he had to deal with, he might very probably have brought them to terms without any bloodshed at all. Smith was likewise guilty, in Clarke's opinion, of great rashness in the manner of making his attack, and especially in making it without having secured a good position to retreat upon ; whereby the Boers were enabled to cut off his communication with the sea. But the courage and skill with which, after the failure of his attack, he defended his camp under every disadvantage against the enemy, made amends for his original errors.

The very candid avowal which Clarke once heard a Boer make to his companion, with respect to the

Caffers, illustrates the *military* character of these 1842
people. I cannot give it in the original Dutch, but
it was to this effect :—" When the Caffer is running
away, then I can shoot with precision, and with
satisfaction ; but when I see him coming towards
me, then my hand gets the palsy ! "

November 5th. I returned from Leeds to London
on the last day of October, having spent five days
with my dear friends Sarah Clarke and her husband ;
and from London to Barton two days ago. Weather
cold and variable, but frequently bright.

Sarah told me some curious things about the
extravagances and improvidence of the manufacturing population of Leeds. In the most flourishing
times of trade, when a working man and his family
might be earning fifty shillings a week, very few
of them saved anything ; the money went as fast as
it was earned, being spent by the men in eating and
drinking, and by the women in dress. It was not
uncommon, I am assured, to see working men buying
turbot and lobsters for their own consumption !—
and the women's extravagance in dress was in proportion to this. Many of the middle classes also
were foolishly expensive in their habits, but their
favourite extravagance consisted in splendid rose-
wood tables, and other costly furniture.

I must not however omit to note down, on the
other hand, what I learned regarding the charity
exercised in these disastrous times by the richer
class of inhabitants of Leeds, and especially the
" Soup Kitchen ", an excellent institution. This

1842 has been established only a few months, but has done great good. The committee who have the management of it cause excellent soup to be served out at the rate of a penny a quart, by which they expect to be £1,000 out of pocket in the year ; and tickets, each entitling the bearer to a quart of soup, are sold in the principal shops, so that any one who is charitably disposed may buy a quantity of these and give them away. It is now, I am told, common for the poor at Leeds to beg for soup tickets instead of money. One advantage of this plan is, that it obviates the risk of throwing away one's money on impostors, Clarke tells me that the soup distributed is of really excellent quality, and highly nourishing.

(He returned to Barton intending to spend the winter at home ; but hearing of the dangerous illness of his brother Edward at Pau, he went off to join him there, and found him convalescent.)

F. J. B.

November 30*th.* At Bordeaux. I little expected, when I wrote the preceding page of this journal, to be *here* at the end of the month. The letter from Edward, which determined me to set off for France, reached us on the 12th. I left Barton on the 14th, but delays of various kinds have caused me to arrive here some days later than I intended, and as the Diligence from hence to Pau goes only thrice a week, I shall not arrive there till the day after to-morrow. I landed at Havre late on the 19th. At Paris I was able to stay only two days and a half,—only enough to convince me that it is the most 1842 amusing place in the world, and to make me heartily sorry to leave it. The sight of the Madeleine and the Bourse makes one feel ashamed of the modern public buildings of London.

December 5th. Bordeaux is a beautiful city, at least the best part of it, for there are plenty of narrow and crooked and dirty streets in the interior. The great theatre is a magnificent building ; and the broad streets and "*places*" near it, especially the Cours de Tourny and the Fosse's de l'Intendance, the handsome parade ground, the long curving line of the quays with their stately buildings, and the superb bridge, are worthy of all praise. I do not think that any of our commercial cities can compete with it in beauty. The streets, however, like those of French towns generally, are ill-paved. The second day of my stay, I had the luck to see a review of the 72nd regiment on this ground,—a pretty sight.

Left Bordeaux for Pau on the evening of the 1st. When daylight came, we were on a sandy plain covered with tall heath and pine trees,—the skirt of the Landes. At sunrise (and a brilliant sunrise it was) I had my first view of the Pyrenees. The open pine woods mixed with heath,—a pleasant, wild and rather pretty sort of country, extend to between Roquefort and Aire ; then comes an enclosed hedge-and-ditch country, not unlike England. At Aire, climbed up a very long and very steep hill.

Between that and Pau, some pretty country, hilly and broken, with much oak coppice, rather remind-

1842 ing me of the neighbourhood of Tunbridge Wells,
though much inferior to that. The day was cloudless, and I had a splendid view of the Pyrenees, their
snowy peaks glittering in the sun. Reached Pau as
it grew dark. I am much pleased with this place.
The town indeed is not well built or handsome, but
it stands well, on the side and brow of a steep hill,
and the view from it is charming. From the terrace
by our hotel we look southwards over the Gave de
Pau, a beautiful little river, clear and rapid, to a
pretty range of steep though not high hills, partly
covered with wood and partly with vines, and dotted
with white houses ; and over these the Pyrenees
tower in all their majesty, forming a glorious background. I scarcely know anywhere a more agreeable
public walk than that in the " Parc " at Pau : the
paths are carried with great taste and judgment
along the sides of a steep wooded bank overlooking
(almost overhanging) the river, and command a
variety of delightful views, in which the cultivated
valley, with farms and villages scattered along it,
the river and bridge, the pretty hills of Jurancon,
and the magnificent Pyrenees, form endless charming
combinations. The Pyrenees, as seen from hence,
are certainly not equal to the Alps, but they are a
beautiful and a grand chain of mountains. The
Pic du Midi de Pau, a remarkable peak, which is
one of the most conspicuous of them, and stands
rather separate from the rest, resembles some of the
Organ Mountains.

* In Brazil.

The castle of Pau, interesting as the birthplace of 1842 Henry the Fourth, is a rather picturesque though not handsome building. It has one great massy square tower, built of thin flat bricks almost like Roman bricks, and three smaller towers, likewise square.

The weather for the last five or six days has been as delicious as it is possible to conceive : a cloudless Italian sun, with a bracing freshness in the mornings and evenings.

December 8th. I am struck with the English character of the vegetation about Pau,—speaking of its general aspect, and as well as one can judge at this time of year. It is utterly unlike the vegetation of the south-east of France : none of the dry aromatic shrubs and dwarfish evergreens which mark that region.

In the hedges I see our common ferns and brambles, plenty of Ruscus aculeatus, Lamium maculatum (replacing our common Lamium album), and a Rubia which appears slightly different from that of Devonshire. The waste shingly tracts between the several channels of the river are covered with Salix purpurea and other low-growing willows, mixed with Juniper. The coppice woods of oak and chestnut which cover a considerable part of the Jurancon hills would no doubt be excellent botanizing ground in the spring and summer. They abound in Mosses ; in particular, I have gathered Leucodon sciuroides, Neckera viticulosa, and Tetraphis pellucida, in the highest perfection, and some delicate Jungermannias. These hills, which abound on the South of the valley

1842 of the Gave de Pau, and seem to extend a long way to the West, are of a material somewhat intermediate between gravel and puddingstone, or it may be called a puddingstone with a soft and almost earthy cement. They are perhaps 700 or 800 feet high, of rounded forms, but very steep on the side towards Pau, as if they had been cut away at some former time by the action of running water. The vineyards which cover much of their surface yield the Jurancon wine, a pleasant white wine, rather strong.

Maize much cultivated in this neighbourhood. Bullock-carts in general use, but much less clumsy and cumbrous than those of Brazil. The country women who come in to market all ride astride. They are mostly coarse-looking, hard-featured, and weather-beaten, as in other parts of France. They appear to be very industrious. The distaff and spindle, which have disappeared from England, or very nearly so, are here seen in constant use.

December 10*th* to 18*th*. In the coppices I find Helle-borus viridis in flower, abundance of the leaves of Pulmonaria officinalis. The great abundance and luxuriant growth of mosses in the hollow ways and woods indicate a moist climate.

It is very curious and pleasing to observe the great variations in the appearance of the Pyrenees accord-ing to the state of the air ; even from the same point of view, they appear one day so distant, another so close at hand, sometimes so high and grand, sometimes so comparatively inconsiderable ; one day almost lost in a bright sunny haze, another grim and

lowering, another so distinct and clear that one 1842
seems to distinguish every rock and cleft and in-
equality of their surface. The largest tower of the
castle of Pau—the heavy square brick tower before
mentioned – was built by Gaston Phœbus, count of
Foix, a great patron of learned men, and of Froissart
among the rest. The walls also are in part of his
time, but subsequently heightened and pierced with
larger windows : the parapet and machecoulis which
surmounted the original wall remain on the present
wall at not more than two-thirds of its height. The
interior of the castle is in process of repair and restoration by the order of Louis Philippe, who has sent down
some fine furniture for it from Versailles. It contains
some interesting historical relics : in particular the
bedstead of the bed in which Henry the Fourth
was born, which is of old dark oak, richly and finely
carved ; the turtle's shell which served as his cradle ;
and his own bedstead. There is also a statue in
white marble of the good king (one of the few kings
of any country who have really deserved that title,)
executed soon after the battle of Ivry, and having
every appearance of a real likeness. In the same
chamber with the cradle is a magnificent vase of
Swedish porphyry, a present from Bernadotte, who is
a native of Pau. The carved chimney-pieces, of the
time of the great Henry, are beautiful. The ceiling
of the great staircase is richly decorated with mouldings of great variety and delicacy, in which frequently occur the interlaced cyphers H. and M., for
Henri and Marguerite. This " Margüerite " was the

1842 celebrated queen of Navarre, the " Marguerite des Marguerites," sister of Francis the First, and authoress of the famous " Nouvelles " in imitation of Boccaccio. Her husband, Henry the Second of Navarre, was the father of Jeanne d'Albret. I was not quite accurate in saying that Pau stands on a hill : it appears so certainly when viewed from the river or the other side of the valley ; but it stands, in fact, on the southern edge of a moderately elevated plain or plateau, which extends for some miles to the north at the same general level as the highest part of the town, and then rises again into a second long terrace or step of land. This formation in successive terraces or platforms seems to be a general character of the country, as may be well seen from the Jurancon hills.

Lescar, anciently the capital of Béarn, and a bishopric, but now a mere village, stands about four miles W. or somewhat N. W. of Pau, on the edge of the same table land. Its church, which is said to have been built as early as A.D. 980, and continued to be a cathedral even till the Revolution, is large but ugly, and in no respect remarkable externally, its interior is noticeable chiefly for the very curious, elaborate, and various sculpture of the capitals of the pillars.

(They left Pau on December 19th, and Charles Bunbury gives an interesting account of the towns he and his brother Edward visited along the South of France before they embarked from Marseilles on the 3rd January, 1843, for Rome ; I shall only quote from

his journal one or two of the places he mentions. 1842
They travelled by Tarbes to Toulouse, the last place
they reached on December 21st, on a cold, raw,
dull morning.)

F. J. B.

December 21st. Toulouse. This is a great city, though
inferior to Bordeaux : it is nearly enclosed between
the Garonne, which is even here a great river, of fine
clear green water, and the famous canal of Languedoc
(or Canal du Midi as it is now called,) which bounds
the town on the East and North, and joins the river
a little below it. There is one bridge, called the
Pont Neuf, over the Garonne, connecting the town
with the suburb of St.Cyprien ; a bridge of ample
size and solid construction, but scarcely handsome ;
it was finished by Louis the Thirteenth. The town
is well supplied with fresh water by several fountains,
and has numerous public walks well laid out, shaded
by plane trees. The Places du Capitole and de
Lafayette, and the part of the city round about
them, are well built and handsome, and improvements appear to be actively going on in many
quarters, but there are also abundance of narrow
filthy streets and wretched dilapidated houses. The
situation of the city, though advantageous, has
nothing picturesque or commanding.

December 22nd. Toulouse. The Capitole, which
contains the Mairie and most of the public offices,
is a large building with a very handsome and imposing front, apparently (the front I mean) of the time

1842 of Louis the Thirteenth or thereabouts. In the first
court of this building, the Duc de Montmorency,
the gallant but unfortunate enemy of Richelieu,
betrayed by the worthless Gaston of Orleans, for
whom he had taken up arms, was beheaded. We
were shown the weapon with which he was beheaded,
a long broad blade somewhat like the bill of the old
English Infantry, but with a short handle. The
great gallery, called the Salle des Hommes Illustres,
is striking in its general effect, though not in perfect
taste ; ranged along the walls are the busts of the
distinguished natives of Toulouse, and under each
bust, on a black marble tablet in a gilt frame, are
inscribed the name and services. There are many
among these " illustres " whom I never happened
to hear of before, but there are also the great lawyer
Cujas, the botanist Lapeyrouse, Riquet, who planned
the canal of Languedoc. The room next to this, in
which the Academy of Toulouse holds its sittings,
contains a monumental figure of Clemence Isaure, a
lady who greatly promoted and encouraged, and long
presided over, the " Jeux Floraux " or poetical contests; for which Toulouse was renowned. These
games are still celebrated yearly, and no doubt tend
to the production of much bad poetry. The figure
of Clemence Isaure was removed hither from her
tomb in the church of La Daurade, and is probably
authentic : she is ungracefully muffled up like a nun,
and the face is not handsome, but the drapery well
executed.

The Museum contains a gallery of pictures, a

considerable collection of Roman sculpture, and a 1842 very rich one of middle-age antiquities.

The church of St. Sernin is very large, very old, (entirely built in the 11th century), and very peculiar in its architecture.

The cathedral (St. Etienne) is also a remarkable church, but of much less antiquity.

December 24th. Carcassonne. The upper or old town, " la Cité " is extremely picturesque and interesting, crowning the top of a steep hill with a complete double enclosure of massy grey walls and towers. Seen from below, from the bridge for instance, it looks like one of the great Welsh castles spread out over a wider space. The walls which extend all round the brow of the hill, following irregularities of its form, are most solidly built of squared stone, battlemented along the top, and pierced with very long loop holes. The towers numerous, round, battlemented and loopholed, those of the inner wall very high. Several of them are very oddly shaped, with a kind of projecting rib or blunt corner, as if they had been once in a pasty state, and had been pinched out on one side. Probably this was to afford a better flanking fire.

There are no machecoulis, and none or hardly any of those pepperbox turrets so often attached to the towers of middle-age fortresses. A castle, probably that of the old vicomtes, embraced between the two lines of wall rises above all the rest with its lofty towers, one of them enormously high, and square, while all the rest are round ; it is of the same con-

1842 struction and evidently of the same period with the rest of the walls, and enhances the picturesque effect of the whole. On one side only the hill slopes down gradually, being connected by a sort of isthmus with a somewhat higher range of hills, and on that side there seems to have been a great ditch, now mostly filled up.

According to the middle-age system of warfare, Carcassonne must have been a very strong place, and it is now a most picturesque one, and very interesting from the complete preservation of its double line of defences.

The old church in this upper town contains a great deal of uncommonly beautiful painted glass, in the windows of the choir, and the great rose windows at the two ends of the transept. One in particular of these rose windows is magnificent. A marble slab in the pavement marks where the savage Simon de Montfort was originally buried. In an outer chapel is the monument of a bishop of Carcassonne who died in 1266 : the carving very rich and elaborate, and in admirable preservation. The new (comparatively new) town of Carcassonne stands in a plain, by the side of the canal ; the river Aude, a clear handsome stream, flows between it and the hills of the old town. The cathedral is a large church, not handsome, but singular for the absence of aisles, and consequently of pillars ; fine painted glass in the choir windows.

The botanic garden of Montpellier, long celebrated, is not of great extent, but seems well arranged and carefully kept up : the conservatories well stocked, 1842 but no great heat; good collection of Cape Irideæ, and of succulent plants ; I saw no Ericas, nor Orchideæ. Nelumbium with its singular fruit fully formed ; the Brasilian Araucaria very flourishing ; five species of Banana ; Bougainvillia in great beauty; a small leaved climbing Ficus entirely covering the wall of a large conservatory with a thick tapestry. In the open air, a very fine tree of the Salisburia, which, being grafted, bears fruit every year ; Cupressus pendula, very handsome ; Magnolia grandiflora, with fruit well formed though not ripe. The common Spruce Fir does not succeed well here, the climate being too hot and dry. The present professor of botany at Montpellier is Delille, author of the " Flora of Egypt."

December 29th. From Montpellier to Nismes. Nismes with its noble amphitheatre and beautiful temple, I had seen before, in 1827, but I found it superior to my recollections.

The Corinthian temple called the Maison Carrée is extremely beautiful, both in the general effect of the whole, and in the details and ornaments ; the work of the frieze, cornice, &c., most rich and elaborate, and at the same time in the purest taste. It is a singular piece of good fortune that it should be so perfectly preserved, really as if it had been kept under a glass case.

December 30th. We took a cabriolet, and drove to the Pont du Gard. The greatest part of the way from Nismes is through the olive-covered plain ; then

1842 we enter rather suddenly among the dry rough limestone hills, clothed with evergreen bushes.   The
ravine between these hills, through which the little
river flows, is pretty in itself, and the majestic
aqueduct, a truly Roman work, makes of the whole
a striking and impressive scene.   This, as well as
Nismes itself, I saw in 1827, my recollection of it was
distinct and vivid.

The Ilex is the predominant plant on these dry
limestone hills, sometimes as a shrub, sometimes as
a tree ; the Box also in plenty ; and on the dry sunny
rocks about the aqueduct, abundance of southern
plants, such as Smilax aspera, Asparagus acutifolius,
the Paliurus, a woolly-leaved Cistus, and great quantities of the garden Thyme (Thymus vulgaris).
Nothing in flower, but a Salvia, and a Euphorbia.
In the way to the Pont du Gard, we had a fine
view of Mount Ventoux, a long ridge gradually
ascending into a point, and covered with snow.

Women's Printing Society, Limited, Great College Street, Westminster, S.W.

1843.

LETTERS.

My Dear Father,

I intended to write to you from Marseilles, 1843 but on our arrival there on the morning of the 3rd, we found that one of the best steamers on the station was to start for Civita Vecchia that same afternoon, and that good berths could be secured in her; so we thought it best to take the opportunity and went on board at once.

We had an interesting tour through the South of France, from Pau, seeing at our leisure several interesting towns, indeed with the exception of Tarbes (which we could not well avoid stopping at) every town on our route presented something of interest. I was very agreeably surprised with Carcassonne; it is a place of which one knows the name in middle-age history, but I was not at all prepared to find it so well worth seeing; the appearance of the upper or old town is extremely curious and picturesque, crowning a steep hill with a complete double enclosure of old, grey battlemented walls and towers, which look (and very likely are) as old as the time of the crusade against the Albigenses. In the old Church of that place, and in the Cathedral of Narbonne, there is the most beautiful painted glass I ever saw.

1843   I was very much pleased with the antiquities of Nîsmes, especially the Maison Carrée, of the beauty of which, as I remember, I was not half sensible enough the first time.   The amphitheatre at Arles, though less perfect on the whole than that of Nîsmes, is very interesting, and being in some particular parts better preserved than that, helps one to make out the structure and arrangement of the whole.   Nearly all these towns in the South of France have handsome and convenient public walks; those of Montpellier are particularly good,— but all these towns too, with the exception of some parts of Toulouse and Nîsmes, are ill-built, with narrow, crooked, dirty streets, and such pavements as it is really a punishment to walk on.   After all I am very glad to have made this tour. and seen such remarkable places; but of all the places I have seen in France, there is not one (with the exception of Paris and Pau) that I have the slightest wish to see again.   I believe the anti-English feeling is as strong in France as ever it was.   It showed itself with great vehemence on the occasion of the bombardment of Barcelona, for they look upon Espartero as completely subservient to England.

We had a fair passage from Marseilles to Civita Vecchia, the weather clear and bright, but very cold; at Leghorn the ground and roofs of the houses, and even the yards of the ships in harbour, were covered with snow.

We spent the greater part of two days at that magnificent place Genoa, and I was quite delighted with its beauty, which surpassed all my recollections 1843 of it. At Leghorn we remained only a few hours, which was lucky, as we had nothing to do there; we passed full in sight of Elba by a fine moonlight, landed at Civita Vecchia yesterday morning; got through the Custom-house with much less delay and annoyance than I expected, and drove post to Rome.

This morning we sallied forth to look for lodgings (for the hotels here are as dear as in London, and appear quite monstrous to us, fresh from the South of France), and by great good fortune, almost at first starting, we fell in with Dr. Somerville, who very kindly assisted us to find comfortable and respectable lodgings, into which we shall move to-morrow. The doctor is grown a good deal thinner than of old, but seems and reports himself to be quite well. Mrs. Somerville and her daughters are looking well, and received us very cordially. I am delighted at the idea of having them for our near neighbours during our stay here, which I hope will be of two months at least.

We have not yet got our letters.

I hope for good accounts of all of you. As for us, I think Edward is as well as ever he was, I have caught cold in coming from Marseilles, but otherwise am very flourishing. I beg you to give my love to Emily and Cecilia.

Believe me

Your very affectionate son,

CHARLES J. F. BUNBURY.

Rome,
February 10th, 1843.

My dear Emily,

1843   Since I last wrote we have had an inundation. On the 4th and three following days so much rain fell, that the wind at the same time blowing *up* the river, the "yellow Tiber" over-flowed ; it rose very nearly to the top of the arches of the Ponte Ste. Angelo, and it was a fine sight to see it rushing down with the rapidity and fury of a mountain torrent. A part of the Corso itself, was flooded, so was nearly the whole of the piazza of the Pantheon, and in the low streets near the river side, the water was so deep that the people were literally going about in boats. The governor of Ste. Angelo told Dr. Somerville that the river had not been so high since 1805. It was at its highest on the 7th : then the wind changed, and though much rain fell the next day, the waters sunk rapidly. For these two days we have had again the bright soft warm weather which Edward execrates, but which to me is delightful.

On the 6th the Alban hills were in great part covered with snow, not merely on their tops, but low down, below Albano itself : but this is nearly all gone. So much gloomy weather has been un-favourable to my intention of seeing pictures, but I have managed to see those of the Vatican, the Farnese and Farnesina galleries, the Borghese (a second time) and the Barberini. Do you remember in this last a Holy Family by Francesco Francia ?

The infant is not good, but the Madonna is 1843 singularly beautiful in features and expression, really worthy of Raphael. I am delighted with Anibal Caracci's frescoes in the Farnese palace, and (which is perhaps wrong) much prefer them to Raphael's in the Farnesina. What a noble portrait is that of Cæsar Borgia, by *Raphael*, in the Borghese palace. In the same gallery there is a St. Dominic, by *Titian*, which I would almost have sworn must be a portrait, yet it cannot be, it is so full of the character of the man, that it seems to breathe the fires of the Inquisition. I daresay you remember, too, an astonishing picture by *Giorgione*, which they call (Lord knows why) David and Saul, so black that scarcely anything can be seen but two magnificent heads, and the gleam of armour actually flashing out of the darkness. But if I go on remarking on particular pictures I shall soon fill my letter, and leave no room for anything else. I am almost ashamed though, of having been five weeks at Rome without seeing the Sistine Chapel. What would Sir Joshua Reynolds have said to me?

Do not, from what I have said in my last letter to you, imagine that I am actually become a convert to the Church of Rome, though I have been living for five weeks close to the College of the Propaganda. I am only so far Romanized, that I rather like the pomp and splendour of their ceremonial, and very decidedly approve of the rich decoration of their Churches, the painting, the gold and marble, the candles, the incense and so forth. The cold, bare,

1843 meagre, shabby look of the white-washed interior of our Churches is deplorable in comparison. A Church may indeed have too much gilding, as in the case of the Gesú, which is, I think, overloaded with ornament, though certainly a fine Church. The most beautiful, next to St. Peter's, that I have yet seen in Italy, is, to my taste, the Annunziata at Genoa.

We see a great deal of the Somervilles, who are very kind to us and very agreeable. We have received much civility also from Sir Frederick Adam and from Mr. and Mrs. Farrar, Lord Colborne's relations. The assassination of poor Mr. Drummond is very shocking. People talk of murders at Rome, Italian bravos and so forth, but the Italians might fairly retort upon us.

Much love to my father and Cissy,

Believe me,

Your's very affectionately,

C. J. F. BUNBURY.

I have for the sake of brevity omitted the detailed and interesting accounts he gives about the Forum and other antiquities, and various discussions on the subject by Bunsen, Nibby and other antiquarians.—(F. J. B.)

JOURNAL.

February. Rome. *Tomb of Eursaycis.* Of all the antiquities of Rome, this, which has been known but a very little while, is by far the strangest and 1843 most indescribable. It is situated just outside the Porta Maggiore, and was, till 1838, buried in the body of a tower belonging to Honorius's wall, so that its existence was unknown until the demolition of that tower in the year aforesaid. It is a building so singular and whimsical in its design, so com- pletely out of all rule, and unlike anything else, as to be very hard to describe. Its ground plan is a trapezoid : one side, that farthest from the wall, or towards the south-east, is broken down ; the other three sides are cased with large regular blocks of travertine, extremely well put together. It consists as it were of three stories, equal in width, the basement is plain, the second story is decorated with a sort of pillars, or rather upright cylinders without bases or capitals, half imbedded in the wall of the building, and varying in number on all the sides : they are most numerous on the north-east side, where there are six of them, in couples, and between every two couples a sort of pilaster as anomalous as the pillars. The third story is the queerest of all : its ornaments (if such they may be called), consist of large hollow cylinders, or wide pipes as it were, of travertine, imbedded in the wall with their open mouths outwards. The number of these, also, is different in each face of the tomb, but in each they are in three rows ; in the face towards Porta Maggiore there are six, two in each row : in the S. W. side there are nine, and in the north-east eleven, one in the uppermost row being

1843 broken away. At the angles of this upper story are pilasters, with curiously ornamented capitals, not referable to any order of architecture, but rather resembling some of those that one sees in buildings of the Byzantine or Lombard style. At the top of all is a frieze covered with small figures in low relief, representing all the process of making bread. And between the second and third stories (as I call them), is another frieze, or border, on which is the inscription informing us to whom the tomb belonged. The good gentleman was not willing that posterity should be in doubt on this point, for the inscription is repeated, with some variations of phrase, on the three sides which remain, and very likely was so on the fourth. On one side the words are :—

M. Marcei. Vergilei Eurysacis. Pistoris. Redemptoris. Apparet.

On another :—

Est. Hoc. Monimentum. Marci. Vergili. Eurysac

The end of the word Eurysacis being broken off. And on the third side the inscription runs thus :—

Est. Hoc. Monimentum. Marcei. Vergilei. Eurysacis. Pistoris. Redemptoris. Apparet.

These inscriptions are extremely well cut in large letters of remarkable neatness, regularity and distinctness. The excellent shape and the deep and sharp cutting of the letters, together with the perfection of the masonry, give reason to suppose that this whimsical monument belongs to an early period, and not to the decline of the empire. The

baker Eurysacis must have been a man of some 1843
wealth and consequence, judging by his tomb. It
appears he was a contractor *(redemptor)*, and not
merely a baker in a small way of trade.

Among the ruins at the time this tomb was cleared
was found another inscription, in letters similar to
those on the tomb, in which somebody (supposed to
be this same baker), informs us that Atistia was his
wife, that she was an excellent woman, and that her
remains are in this *bread basket*.

March 3rd. Rome. We took advantage of this
fine day, after so much rain, to explore the Palatine.
This hill is partly occupied by the house and
gardens of an Englishman, Mr. Mills,—partly by the
Farnese gardens, belonging to the crown of Naples,
—and partly by those belonging to the English
College. These last, which cover the south-western
part of the hill, contain by far the most important
and picturesque ruins. From the top of these ruins
we enjoyed delicious views: in one direction, the
old battlemented towers of the Porta S. Paolo, with
the sharp pyramid of Caius Cestius beside it, the
low, green undulating hills of the Campagna beyond,
the light catching here and there on the windings of
the Tiber, and finally the flat line of the sea-coast,
with single towers scattered along it from distance to
distance ; in another direction, first the grand mass
of Caracalla's baths, then the towers of the Porta
S. Sebastiano, beyond them the tomb of Cecilia
Metella gleaming white in the sunshine, the green
Campagna, and the mountains of Albano, sometimes

1843 cast into deep shade by the passing clouds, sometimes glowing with sunshine and sparkling with
their clusters of white houses. Looking another
way, again, we saw a part of the modern and
crowded city, the river, the Janiculum crowned with
the beautiful pines of the Villa Pamfili, and the
glory of modern Rome, the dome of St. Peter's.   To
all this, a noble foreground was formed by the grand
arches and vast masses of masonry which remain of
the Cæsars' palace, magnificent in ruins, and rendered more beautiful by a rich growth of wild
flowers and evergreen shrubs, which clothe, without
too much concealing them.

Of the ancient remains on the Palatine, though
the greatest and most imposing masses are, as I
have said, in the gardens of the English College,
towards the church of S. Gregorio, and towards the
Aventine, the most curious perhaps are the sub-
terranean chambers called the Baths of Livia, which
are in the Farnese gardens.   The vaults of these
chambers retain their fine stucco, covered with very
beautiful and delicate arabesques, the colours and
gilding very well preserved,—more perfect, or at
least more easily visible, than those under the baths
of Titus.

Profound antiquarians and learned architects may
perhaps be able to trace or to imagine something of
the plan of these ruins ; but to me they appear all
confusion,—striking to the eye, interesting to the
imagination, but unintelligible as to their design and
arrangement.   One sees vast masses of brick wall,

long rows of lofty arches, numberless vaulted 1843
chambers more or less complete, some of the vaults
enriched with deep cassettoni, passages which at
present certainly lead to nothing, and in short, a
magnificent but unintelligible maze of building.
There are not, however, (as far as we could see),
any halls comparable in size to those in Caracalla's
Baths. Portions of stucco remain in several places
on the walls and ceilings, and in one or two
instances retain something of their colours.

March 5th. I walked out to the Ponte Lamentano
(Pons Nomentanus) over the Anio, and had a most
beautiful view of the Alban mountains, the snowy
Apennines, Soracte, and the hills farther to the
north-west beyond the Lake of Bracciano. Much
snow again on the Alban mountains, especially on
the Monte Cavo and on the high plain known as the
Campo di Annibale. I could not distinguish any
snow on Soracte.

One sees very well from hence the general
character of the Campagna ; it is not a perfect flat,
nor yet a hilly country, but a slightly undulated
plain cut in various directions by shallow ravines
and valleys, which divide it into a number of tabular
eminences with steep and often precipitous sides ;
the whole covered with short grass, except here and
there on the escarpments. The ruined tombs and
scattered towers dotting the surface of these lonely
plains, have a very singular effect.

I found few flowers, much fewer than I expected,
but plenty of large bluish-purple Anemones (A.
coronaria) already in blossom on the Campagna.

1843 *March* 15*th.* Perhaps the most beautiful of all the views of Rome is that from S. Pietro in Montorio on the Janiculum. The variety of the scenery included in this view is astonishing; on one side the irregular crowded mass of the modern city, with the cupolas and towers of its innumerable churches; on another, the wide extent of gardens and vine-yards which, interrupted only by a few churches and convents and scattered houses, occupy all the southern and eastern part of Rome, now gay with Almond blossoms, and with the bright, fresh green of spring; the ruins and the Cypresses of the Pala-tine; the three magnificent arches of the Temple of Peace; and the glorious Coliseum rising in its majesty above all surrounding objects, and seeming to command the ancient city as St. Peter's commands the modern.

The Campagna beyond the gates of S. Sebastian's and S. Paolo, towards, the Alban mountains, and towards Ostia, is seen fully as well from hence as from the Palatine, and to the north one sees a long reach of the valley of the Tiber above Rome, and the open country extending to the hills of Bracciano.

The solitary mountain of Soracte, rising so abruptly out of the plain, with its serrated crest, is a very remarkable object in the distant view.

San Clemente. One of the oldest and most curious churches in Rome. It is on the left hand as you go to S. Giovanni Laterano from the Coliseum; the outside, as usual in the old churches here is not conspicuous. It has, like the other old churches which

are called Basilicas, a rich mosaic pavement of red 1843
and green porphyry and white marble ; no transept ;
the aisles divided from the nave by columns
taken from ancient buildings, dissimilar among
themselves in form and material; a semi-circular
and vaulted *apsis* or *tribune* at the end behind the
high altar, lined with mosaics in the stiff and gaudy
Byzantine style, on a gold ground. These are the
common characteristics of the old Roman churches.
To proceed to what is more particularly remarkable
in this one.—A part of the nave, near the altar, is
raised above the level of the rest, and enclosed by a
sort of low wall of white marble, richly inlaid with
mosaics, and with porphyry and other coloured
stones. On either side of it are the pulpits of
marble, called Ambones, from which in old times
the epistles and gospels were read to the congrega-
tion. The end of the Church, behind the high altar,
is raised somewhat higher yet, and separated from
the rest by a low screen of marble, beautifully
wrought in a variety of patterns, in compartments.
Quite at the extremity in the tribune, is a marble
chair, which used to be occupied by the Bishop,
and on either side of it seats of the same material
(segments of a circle) for his clergy.

One chapel in this church contains some interest-
Frescoes by Masaccio, representing the Crucifixion
and some of the events of the life of St. Catherine.
They have, of course something of the usual stiffness
of those early painters,* but at the same time, a

* Masaccio was born 1401, and died 1442.

1843 fine simplicity and quiet dignity; several of the heads are very noble in character and expression; and in the Crucifixion, there are two or three figures which are quite worthy of Raphæl.

Circus of Caracalla. The antiquaries now call this the Circus of Romulus, the son of Maxentius, being of opinion that the style of construction of its walls indicates a much later period than that of Caracalla. It is, I believe, the only ancient Circus which remains at all entire. Its form is a much elongated parallelogram, rounded off at one end: the low wall enclosing it remains complete, but there are no distinct traces of the seats: the end which is not rounded is flanked by two large towers, and at the other there is a high arch; the Spina, a low platform of masonry, runs along the middle, approaching, however, much nearer to the rounded end than to the other, and is bounded at each end by a low building containing a small vaulted chamber. All these walls are built of alternate layers of thin flat tiles and irregular pieces of tufa with a great deal of cement,—a construction certainly very different from that of Caracalla's Baths, and one sees no casing of fine brickwork as in that instance, The inscription to Romulus, the son of Maxentius, which is now put up in the archway, was discovered when the interior of the Circus was cleared out, by the order of the Duke of Bracciano, on whose property it is situated. A large rectangular enclosure of high brick walls, of similar construction to the walls of the Circus, is

commonly known as the "Scuderies" or stables of
the Circus, to which it is adjacent: but Nibby
supposes it to be the enclosure of a temple dedicated
to the same Romulus, the son of Maxentius. Near
this stands what is called the *Temple of Bacchus*;
a rectangular temple of moderate size, but very
high in proportion, built entirely of brick, except
four columns of grey marble in front, supporting a
low entablature of the same material. What is very
peculiar, is, that the pediment does not rest
immediately on this entablature, but above the
entablature there is a considerable height of wall,
as if an upper story, and then a frieze and
cornice of brick, highly ornamented with very rich
and handsome mouldings. The brickwork of the
whole building is very good, and has that warm,
soft, tawny red colour, which is often seen in the
brick buildings of ancient Rome, and which is
peculiarly agreeable to the eye. The name of
Temple of Bacchus is supposed to be justified by an
altar dedicated to that god, which was found in
excavations underneath it.

This Temple is prettily situated, nearly on the
brow of a smooth green hill, sprinkled with trees,
commanding an extensive view of the Campagna;
near it is another hill crowned with a grove of Ilex,
and in the valley below it is the so-called *Grotto of*
Egeria. The Stanzas in "Childe Harold" are
calculated to give a much exaggerated idea of this
nymphæum. The quiet secluded grassy valley in
which it is situated, and the general character of the

1843 scenery about it, are indeed very pleasing, but the spring, though clear and bright, is very small, and the grotto much more remarkable for what has been said or sung about it than for anything in itself.

From the grotto we had a very pleasant walk across the Campagna to the magnificent arches of *Claudius's aqueduct.* These aqueducts, which are seen stretching in so many directions, and for so many miles over the open country, are certainly among the noblest works that remain of the Roman Empire, and nothing contributes more to the picturesque character of the country around Rome.

The Claudian aqueduct is built of great square blocks of peperino, while most of the others are of brick.

In going back to the Naples road, we passed a Temple which the guides have christened the Temple of Fortuna Muliebris, supposing it to be the one erected when Coriolanus turned back ; but it is as uncertain as most of these remains.

It is built of brick, and has a great general resemblance to the temple of Bacchus, to which however it is inferior in the execution of the details.

The aspect of this part of the Campagna, especially in such clear and brilliant weather as we have now (March 18), is very impressive, and with all its loneliness and desolation, even beautiful. In all directions are the remains of ancient buildings, memorials of the mighty Romans of old,—gigantic aqueducts, tombs without a name, fragments of temples or villas: and these, scattered over the wide 1843 expanse of smooth green turf where there is scarcely a cottage or a living man to be seen, appear to claim to themselves the dominion of the wilderness. Farther off, in contrast to these, are the bright white houses so thickly dotted over the Alban mountains: and in the opposite direction, the cupolas and towers of Rome.

[He also describes, in detail, the principal pictures in the following Palaces of Rome. — The Borghese, Barberini, Rospigliosi, Farnese, Farnesina, Colonna, Doria. These descriptions I shall place in Appendix at the end of the volume.]—F. J. B.

LETTERS.

Naples,
April 15th, 1843.

My Dear Emily,

I arrived here on the 13th, and yesterday received my father's letter of the 28th March, for which I beg you to thank him in my name. I write to you this time, because I owe you a letter.

Before leaving Rome, Edward and I made an excursion of three days to the Alban mountains, explored the neighbourhood of the two lakes, ascended Monte Cavo, and visited Frascati and Tusculum. This was very enjoyable. I was delighted with the quantities of beautiful wild flowers in the woods:—such profusion of the sky-blue wood Anemone and the crimson Cyclamen and the Lungwort, besides many that I had never seen

D D

1843 before: and on Monte Cavo I found the yellow
Anemone, (ranunculoides), the beautiful little Scilla
bifolia, the Fumaria bulbosa in profusion, the white
Narcissus with a scarlet edge to its cup, (poeticus)
and other novelties.   Certainly the spring flowers of
Italy are very charming.   For a fortnight or so
before I left Rome, the Campagna and all the grass
plats of the villas were full of the Spider Orchis and
the bright red Orchis papilionacea, and new kinds
were coming out every day, so that I could hardly
keep my list up to the *courant* of vegetation.   The
remains of Tusculum are very interesting, especially
the theatre, which has but very lately been entirely
cleared out, and is remarkably perfect : very nearly
as perfect, Edward says, as those at Pompeii : the
seats all in complete preservation, with the ranges
of steps dividing them, and the walls and sub-
divisions of the scena nearly entire.   There are also
great remains of the city walls, and the pavement of
several of the streets and of part of the Forum,
and we saw the very road up which Mamilius's
horse ran from the battle of Regillus.   I have no
doubt that by farther excavations a great deal more
of the plan of the ancient town might be discovered.
But the most perfect piece of ancient pavement I
have seen anywhere, is on the ascent of Monte
Cavo : there is a great length of it, and the great
polygonal stones are as even and as neatly joined as
if they had been laid down yesterday.

I found Rome so delightful after the fine weather
set in, that I was very unwilling to leave it, even for

Naples. I did leave it, however, on Monday last, 1843 and came hither in four days by vetturino, sleeping at Cisterna, Terracina and Sant Agata.

Edward remains behind, wishing to see the ceremonies of Easter, and then to look for the remains of the ancient Latin cities.

The weather, which changed just before I left Rome, was not favourable to my journey, yet I was very much pleased with Mola di Gaeta, and with much of the country between that and Naples. Terracina is a very picturesque place, but what a miserable and ruffianly-looking population ! The approach to Naples is, to be sure, a striking contrast to that of Rome. I am here in an excellent hotel, the Gran Bretagna, the master of which (Buonaccorsi) says he was once in my father's service, I believe as a courier. The situation is delightful : I look directly down from my window on the public garden, all rosy with the blossoms of the Judas trees, and over that to the bay, which at this moment is all speckled with fishing boats.

Yesterday I could not judge well of the scenery, for there was so thick a fog all day, that Vesuvius was but barely visible, and though to-day is very fine, there is still a heavy bank of fog all over the bay, the tops of the mountains just appearing above it.

In consequence of the festa (yesterday being Good Friday), I have not yet been able to see the Museum, but I perceive that the festa days do not prevent the Neapolitans from working in their

1843 vocation, for I had my pocket picked before I had been here 24 hours.   They seem to be *smart* people, as the Yankees would say.

I hope the Rickardses* are well.   I always bear them in mind when I am collecting plants, and have got plenty of specimens for them.

I am very sorry to hear that John† has been so dangerously ill, but I trust he has recovered effectually.

I hope you saw the comet, which was a mighty fine fellow, and made a great sensation at Rome, where there was much marvel at the length of its tail, and much speculation as to its effects on the weather.   I was mistaken in saying, in my last letter to my father, that nobody had seen its head: Mrs. Somerville saw it, and I understand the astronomers at the Observatory got satisfactory observations of it.   Everybody at Rome was quoting Mrs. Somerville at random, and making her answerable for many things she never dreamt of.

Yours very affectionately,
C. J. F. BUNBURY.

Naples,
April 30th, 1843.

My dear Father,

I thank you much for your letter of the 28th March, which I received on my arrival here, and I am very much obliged to you for your offer of farther assistance in money to complete my tour in

---

* The Rev Samuel Rickards, rector of Stowlangtoft, and his wife and daughter.

† John Napier.

Sicily. I believe that what I have will enable me to 1843 make out that tour satisfactorily, and carry me to Malta, but not farther. I shall therefore be very glad if you will send to meet me at Malta a letter of credit for a sum sufficient to carry me home. Pray let me know also what is Mr. Sconce's* office there.

I wrote to Emily a couple of days after my arrival here, and gave her an account of our excursion to Albano, Monte Cavo, and Tusculum, which was a very agreeable one. I have been enchanted with the museum here, which I have visited almost every day since it opened : it is altogether the most interesting collection I ever saw, and I should never be tired of it. The infinite beauty and variety of the bronzes found in Herculaneum and Pompeii give one a wonderful idea of the exquisite taste and universal refinement of luxury of the ancients, when we find such treasures in two provincial towns, by no means of the first class. Some of the statues are beyond all praise : the Aristides (or Æschines) as a perfect and living representation, not of anything ideal or superhuman, but of a wise, dignified, and thoughtful man, is as superlative as the Belvedere Apollo is in the ideal class. The so-called Psyche, from Capua, is the most lovely face I ever saw, either in art or nature. The Venus is very beautiful,—not her face, but her figure ; I never saw marble with so perfectly the appearance of flesh. The Agrippina is an admirable statue, and

* The father of his sister-in-law, Mrs. Hanmer Bunbury.

1843 the air of melancholy and dejection which charac-
terizes it, makes it perhaps more interesting thau
the Agrippina of the Capitol, which is equally
excellent for the graceful dignity and ease of its
attitude.   There is no excellence, I think, more
striking in ancient sculpture than these,—the simplicity, ease, and quiet unaffected dignity and grace.
As for the Toro Farnese, it never took my fancy
much in the copies or castes I had seen of it, nor, I
confess, am I much charmed with the original.

The ancient pictures, of which they have now
a great number, are extremely curious and interesting for the information they give us respecting
the methods of the ancients in that art, as well
as their choice and treatment of subjects ; and
some of them are of much more striking merit than
what I had expected to see.   Blessings on the
eruption of Vesuvius, which by covering up those
towns preserved to us so much curious information
about those wonderful nations of antiquity, which,
with all their faults, have so powerful an influence
over our minds.   In the collection of modern
pictures, I was much struck with Titian's portrait
of Paul the Third and his nephew, the magnificent
portrait which they call Christopher Columbus,
Schidone's Charity, Domenichino's Guardian Angel,
and (as a very vigorous and clever, though not
an agreeable picture), Spagnoletto's Silenus.

I have as yet seen scarcely anything but the
museum, which indeed is enough to occupy one for
a good while.

I went one day to Pozzuoli with my friends the 1843 Farrers, and saw the Temple of Serapis, the Amphitheatre, and the Solfatara. I have not heard from Edward since we parted : I am afraid I shall not persuade him to join me in my Sicilian tour. It would be very pleasant if he would, for I find that I understand and appreciate antiquities much better when I see them with him, and one feels the disagreeables of rough travelling far less when one has an agreeable companion. But his mind seemed to be bent on the Lake of Fucinus and the Samnite Apennines.

I hope Emily and Cissy are in better health than at the time of your last letters. Pray give my love to them, and believe me

Your most affectionate Son,

CHARLES J. F. BUNBURY.

Naples,
May 16th, 1843.

My dear Father,

I am just returned from an excursion to Pæstum, which I enjoyed as much as I could *alone*, but I wished very much that you and Emily had been with me. I have been much disappointed by Edward's giving up his plan of joining me here, as in visiting antiquities in particular, I feel the want of an intelligent companion, and one who understands the subject so well, and takes so much interest in it as he does. I have missed him much. Since the Farrers went I have been entirely alone. The weather has been delightful, and I saw the

1843 great temples to full advantage,—and magnificent they are indeed. The Temple of Neptune is one of the most sublime of human works that I ever looked upon,—it has an effect of majesty and grandeur far beyond its real size. Their situation too is most impressive. I was delighted with the beauty of the valley of La Cava and the coast of Salerno, indeed it is altogether a lovely country ; the luxuriant beauty and richness of the valleys and plains, with their vines and fig trees and corn, and the bright white houses, and the convents perched so picturesquely on the hills, combine so admirably with the great wild frowning mountains. Since I left Rio de Janeiro, I have seen nothing like the exuberant fertility and garden-like beauty of this country. I was at Castellamare on a great festa, when the streets were crowded with people in their gayest dresses, and the scene was exceedingly animated and pretty. I was much pleased with that place. In my way back I took the oppportunity of going through Pompeii for the third time, so I think I know it well now ; and I never saw a place that pleased and interested me more. It had been one of my greatest objects in coming to Naples, and certainly it has not disappointed my expectations. It is singularly curious and impressive, though one cannot help wishing that more of the paintings and decorations had been left in their original places. Some few *of* the pictures which have been left under the shelter of a roof or shed, are in as good condition as most of those

in the museum, so it seems as if it would have been 1843
possible to preserve them from decay without removing them. Still, as it is, there is enough to
show the taste and fancy of the ancients in decora-
ting their houses, and one is surprised at the beauty
and variety and delicacy of the arabesques and
fanciful devices lavished everywhere, and at the
care and taste and cost with which even the smallest
chambers are ornamented. It is quite curious to
see the elaborate paintings, and in some cases rich
mosaics lavished on bedrooms as small as monks'
cells, and which for the most part had not even
a window. The Pompeians certainly were contented
with wonderfully small sleeping places. I had
heard so much of the narrowness of the streets, that
I was not struck with it ; they do not appear to me
narrower than those in most of the second-rate towns
of France, and they show a degree of attention
to the public convenience which one seldom sees
either in France or Italy, in being provided with
footways. The public baths, which are beautifully
preserved, are very interesting and satisfactory,
making us so much better acquainted than anything
else can with the plan of those establishments so
important among the ancients ; for the baths at
Rome, though most picturesque and magnificent
ruins, are very unintelligible. Probably we must not
take Pompeii to be exactly a miniature of Rome,
but must allow for various differences in applying to
the latter the ideas derived from the former : we
know for instance that at Rome the houses were

1843 very high, while at Pompeii they were evidently very low; but still this place gives one a most gratifying insight into the habits and tastes and domestic life of the great nation. As the Americans would say, we *realize* the Romans so much better after having seen Pompeii.

Herculaneum is very unsatisfactory to see, it is just like going down into a mine, which is a thing I detest. The coast of Baia is exceedingly pretty, and though few of the antiquities there are separately worth much, it is striking to see the great extent and continuity of the ruins lining the whole shore, and agreeing with the accounts we read of its former populousness. But I was unpleasantly struck with the miserable and poverty-stricken appearance of the present inhabitants, so strongly contrasted with the fertility of the country. They are as beggarly, and appear almost as wretched, as in any part of Ireland.

I mean to go up Vesuvius to-morrow. I propose to set off for Palermo by the steamer on Monday next, the 22nd, thence to make the tour of the principal curiosities of the island, and expect to be at Messina in time for the steamer which goes to Malta about the 16th of next month. I wish indeed I could have your company in this tour. I hope I shall hear from you at Malta.

It was not till the other day that I saw the account of Sir Charles Napier's victory in the Scinde country: it is splendid! and I wish Emily joy, as I am sure she must be proud of it.

I hope there are satisfactory accounts of John. I 1843 am anxious to hear again from you concerning Cissy, as your last account of her was not at all pleasant.

Pray give my love to her and Emily,

Believe me ever;

Your very affectionate son,

C J. F. BUNBURY.

P.S.—I am sending off a box containing my dried plants, and some books of prints I have bought here, and I direct it to you, and I hope you will have it taken care of for me.

JOURNAL.

May 23rd. Palermo. I am much struck with the beauty of this place. I have seldom seen a bay of more graceful outline, nor headlands and rocky points more picturesquely grouped and more finely broken and varied in their forms than those which enclose this bay. The mountains which sweep round the city and its plain in a fine amphitheatrical curve, are still more barren, bare and grey, than those of Genoa, but much bolder and more picturesque in outline. This scenery wants the splendid vegetation and rich green of the topics, but it wants little else to render it perfectly beautiful.

I had a very pleasant drive to the villa of La Bagaria, eight or nine miles from the city: the road

1843 at first along the sea-shore, afterwards through the rich and beautiful plain which is enclosed between the sea and the mountains. This plain, irrigated and highly cultivated, displays very luxuriant vegetation and a rich variety of crops : olives, vines, corn, mulberry and fig trees, the tall Italian reed, and the prickly pear, form a beautiful mixture. The very great abundance of the prickly pear is a remarkable feature in the landscape : the roads and fields are everywhere bordered with it, and it grows as large, and with a trunk as thick and woody, as ever I saw it in S. America. Its yellow flowers are just now coming out. The American Aloe, too, is seen in very many places along the roadside, but, like the Cactus, evidently planted. At La Bagaria, ascend one of the abrupt limestone rocks rising out of the plain, from which there is a most beautiful view of the Bay of Palermo, another bay divided from it by the high rocky headlands of Zafferana, and the fine mountainous coast to the eastward of it, the rich plain of Palermo, and the amphitheatre of rough stony mountains around it.

In the same neighbourhood are villas belonging to Prince Palagonia and Prince Butera, both delightfully situated : the former used to be famous for its monsters carved in stone, of which few now remain. The only thing remarkable I saw in it was a saloon, of which the ceiling is all looking-glass, and the walls lined with glass coloured in imitation of agate. Prince Butera's villa is another specimen of whim, fitted up in imitation of a convent with dressed-up wax figures, as large as life, in every room and in 1843 every passage.

The Botanic Garden I saw very hastily, but was pleased with the vigorous and thriving appearance of the plants, of which it seems to have a very considerable variety, though I did not happen to notice anything that was altogether new to me.

The Villa Giulia, or public garden, called also La Flora, is, I think, superior to the Villa Reale at Naples, it has more variety and more shade. Its avenues of orange and lemon trees, and hedges of China roses, put me in mind of Rio de Janeiro, as indeed do various things at Palermo. The vines in this neighbourhood are trained rather low, not festooned from tree to tree, so that they do not produce as rich and beautiful an effect as in the Campagna of Naples. The grey of the olive, too, perhaps prevails rather too much in the landscape.

The cattle here are very different from what I have been used to see in Italy, and far less handsome. They are almost all brown or red, smaller than the grey Italian cattle, but with immense horns, as long as those of the Caffer oxen, and these almost upright.

May 24th A broiling day,—the first very hot one I have felt this year.

Ascended the Monte Pellegrino, a singularly, rugged, abrupt, and craggy limestone hill, quite insulated from the rest of the mountains, standing out boldly as if for a barrier between the fertile plain and the sea, and sheltering the harbour of

1843 Palermo on the west.   Its height is said to be
     1960 feet.   The grotto of Santa Rosalia, in which,
     according to tradition, she died after spending many
     years in the practice of continual austerity and
     self-torture, is in the side of this hill farthest from
     Palermo, some way below the summit and at the
     foot of a great cliff of limestone.   This cavern,
     converted into a chapel, is partly lined with
     fine marbles, but much of the rough natural rock
     remains exposed, and it has a curious effect : thus
     seeing the dark, rugged, damp walls and roof of the
     cave in such close juxtaposition with the profuse
     and gaudy decorations which the Sicilians and
     southern Italians love to bestow on their churches.
     The ascent to the grotto from the plain is by a
     rough stony path, not steep or difficult, yet fatiguing
     enough in such hot weather as this.

     Is must, however, have been constructed at con-
     siderable expense and trouble, for it is supported
     for a long way together on arches.   It commands
     admirable views of Palermo and its bay and
     neighbourhood.   The city, standing on level ground,
     makes little show when seen from the sea or from
     its own plain, but has a very handsome appearance
     when thus viewed from a height.

     This Monte Pellegrino, and all the mountains
     near it, are most strangely bare, stony and sun-
     burnt, as much so as any of those in the interior of
     the Cape colony ; not a tree not a bush above a
     foot high, and even the herbs growing thin and
     scattered in the interstices of the loose parched

stones. A far as I could observe, there seemed to 1843 be no great variety of plants, but those that there were, rather peculiar. I noticed a Biscutella, a Paronychia, a pretty Convolvulus with digitated leaves, plenty of a delicate little blue-flowered Sempervivum, and a curious leguminous plant with inflated calyces,—I suppose an Anthyllis: abundance also of Ægilops ovata and Echium violaceum.

Descending from this mountain, I went to the Royal Villa of La Favorita, situated in the plain at its foot :—little worth seeing. Thence to the villa of the Duke of Serra di Falco, which is very pretty, shady and pleasant, but debased by several of those silly conceits and childish tricks which seem to be in favour with the Palermitan nobles,—wax-work hermits that start up and grin at you when you approach their hermitages, and unexpected jets of water that play upon you from concealed fountains. At the Favorita, the cicerone showed me a cock pheasant as something rare and admirable.

May 25th. Palermo. The Royal Chapel is very remarkable. It has a great resemblance in many respects to the oldest Basilicas at Rome, but its arches are rather pointed. It has no transept :— a very lofty nave and aisles, divided by simple columns of various materials, some fluted and some plain, on which rest the aforesaid arches ; the whole of the space above and between these arches, as well as the roof both of the nave and aisles, and the upper half of the walls of the aisles are covered with mosaics on a gold ground, in the stiff and gaudy

1843 Byzantine style, like those in the tribunes of the old churches of Rome. The pavement is a very fine specimen of the rich tesselated work of porphyry and marble, which is a characteristic of those old Basilicas ; the side walls up to a certain height, of white marble inlaid with mosaic ; and at the upper end of the nave, on the right hand side, is a magnificent inlaid marble pulpit, like one of the Ambones in S. Lorenzo at Rome.

I was shown the apartments of the Royal Palace, but thought them little worth seeing. The Cathedral is, outwardly, a fine and striking specimen of Norman or Early Gothic architecture, but the interior, unfortunately, is entirely modernized. It contains some magnificent sarcophagi of porphyry. extremely well wrought, and placed under canopies supported by columns of the same material.

The Via di Toledo, the principal street of Palermo, though not a match for its name-sake at Naples, is a long and fine street. Its very high solidly-built houses, with their far-projecting balconies and the singular latticed or grated galleries running along the whole front of the upper story of many of them, have a certain dignity and solemnity of character corresponding with my idea of Spanish towns. Shops occupy the ground floor of most of the houses.

The architecture of the houses looking on the Marina is in a lighter, gayer, and somewhat fanciful style. This same Marina is a very pleasant walk after the heat of the day is over, and commands a delightful view of the bay. I lounged for some time 1843 in the public garden, which, as the day was a festa, was pretty full of people; but I saw few pretty women. My favourable opinion of the " Flora" was however confirmed. The shade, the abundant vigorous foliage, and the profusion of flowers make it very agreeable. The luxuriance of vegetation here, wherever the soil is sufficiently watered, is almost worthy of the tropics, and makes a singular contrast with the extreme barrenness and nakedness of the neighbouring mountains.

Some writers speak of the abundance of Palm trees (date palms) at Palermo. I was not struck with it. There are some uncommonly tall trees of that kind in the Botanic Garden, two or three in the " Flora," one or two fine ones in the villa of the Duke of Serra di Falco ; these are all I have happened to see. They do not make so much show here as at Terracina. In the public garden, there are some of the finest Plane trees I ever saw, Beadtrees *(Melia)* of great size and now coming into full blossom, abundance of Judas trees loaded with their dark red seedpods, a flourishing Bamboo (in the open ground) and Orange and Lemon trees *ad libitum.*

Palermo is an amazingly noisy city on a festa— not much less noisy than Rio de Janeiro.

The Hotel d'Albion is a very good inn.

May 26*th.* Started from Palermo about 7 a.m. (an hour later than I intended) with two saddle mules and two baggage ditto. From Palermo at least half way to Monreale is a continued suburb ;

1843 thence a moderate ascent to Monreale, an old town situated low down on the side of a mountain, making a very fine show at a distance, but very vile when one is in it. The slope below it and the adjacent plain for some distance towards Palermo look like a continued wood of olive, orange, mulberry and fig tree.

The Cathedral of Monreale, a very large, old and curious church, much in the style of the Royal Chapel at Palermo ; all the Old and New Testament in gaudy mosaic on the walls and above the arches of the nave. Thence, a gradual ascent for several miles along the flank of a limestone mountain ; a beautiful rich green valley on the left and rough craggy mountains beyond. The road and fields bordered with Cactus and Aloe : the latter now shooting up its young flower-stems, like giant asparagus. From the head of the valley the road ascends zigzag some way farther, leaving cultivation nearly, but not entirely, behind ; then descends several miles, a narrow pass between limestone mountains, excessively steep and rugged. A very great variety of pretty flowers among the rocks and by the roadside, especially Vetches, Trefoils, Marygolds ; no trees and scarcely a bush on the mountains. Descend into another fertile plain (but less beautiful than that of Palermo) between these mountains and the sea, and arrive at La Sala de Partinico (so my guide called it), a town of some size, very far from handsome, but finely situated near the underfalls of the mountains. It is 19 miles

from Palermo. Rest here an hour at a miserable 1843
publichouse ; then go on 13 miles farther to Alcamo,
over cultivated but almost treeless hills, with fine
mountains on the left and in front. The beauty
and abundance of the wild flowers are very striking ;
but I am too sleepy to enumerate any.

Alcamo, with its numerous churches and convents
ranged along the ridge of a hill, looks very picturesque as we approach it. The inn is very
Brasilian.

Thus far there is a tolerably good carriage road,
and much frequented. I walked at least two-thirds
of this day's journey, finding the mule's pace more
fatiguing.

May 27th. Alcamo is by no means an ill-looking
town : the principal street long, straight, reasonably
wide, and not ill paved ; but it seems to have an
amazingly large proportion of churches and convents. The principal church large and handsome.

Start at half-past six, proceed a little way
along the high road ; then turn to the right and
proceed to Segesta by a mule path, over round, tame
hills covered with corn and thistles. Cross the
stony bed of a stream filled with wild thickets ;
here, for the first time, I see the Palmetta and
Oleander growing wild. The pretty Convolvulus
tricolor abundant among the corn and on the fallow
lands, but smaller than in gardens. After nearly
three hours riding come to bolder and more
rocky hills ; and on one of these, between two

1843 higher hills, alone and majestic, stands the Temple
of Segesta.

It is truly a magnificent building, though perhaps
not quite so impressive as the Temple of Neptune
at Pæstum. Its columns, 36 in number, differ in
form and proportion from those of Pæstum ; they
are not so thick for their height, less bulging in the
lower part, less rapidly narrowing upwards, and
their capitals much less broad. These columns are
not fluted. There is no trace of columns in the
interior, as at Pæstum, nor any part of the interior
raised above the level of the rest. Entablature and
both pediments remain entire ; but the roof gone, as
usual. The whole is built of a very porous calcareous stone, apparently a sort of travertine, but not
a congeries of petrified vegetable fragments like the
Pæstum stone.

Some little way from the temple, nearly at the top
of a very steep hill, is the Theatre, with its seats
and the flights of steps intersecting them in fine
preservation. It is not altogether hewn in the side
of the hill, though advantage has been taken of the
slope, so that the height of the external wall is much
less in the middle part of the circumference than
towards the ends of the arc. This wall is built of
long, unequal and rather irregular (though in
general nearly rectangular) blocks of limestone,
without cement, with a quantity of small stones here
and there rammed into the interstices. The seats
are built of more regular blocks of the same stone.

The Temple and Theatre of Segesta are now

inhabited by great numbers of hawks, crows and 1843 jackdaws.

Leaving this, we wound along the base of some rough rocky limestone hills, having a noble view of the temple as we departed from it. The Palmetto (Chamærops humilis) grows in great quantities among the rocks; its dark green clumps of stiff sword-blade-like leaflets, scarcely more than two feet high, have a singular appearance.

See the town of Calatafimi at a few miles distance on our left. Rejoin the main road and descend into a rich valley. Heat suffocating and no shade. Pass Salermi (?), a curious old grey town, most picturesquely placed on a steep hill, with an old tower at the top. Thence to Castle Vetrano, a wearisome journey over long, dull, tame clay hills, without a tree, a hedge, or even a bush—nothing but corn and thistles. It put me in mind of what I have read of the interior of Spain. With all this extent of cultivation, one sees very few houses; it would seem as if all the population were gathered into towns. We met, however, great numbers of men and mules going to a fair which was held at the aforesaid picturesque grey town.

In more than one place I noticed great rocks of glittering flaky crystalline gypsum (selenite).

Arrive at Castel Vetrano very much tired, the day having been tremendously hot, and the journey of more than twelve hours, including the exploration of Segesta. The distance is called 35 miles. The Inn at Castel Vetrano is some degrees less miserable than at Alcamo.

1843 *May 28th.* From Castel Vetrano to the sea-shore, about eight miles and much pleasanter country, enclosed and well cultivated, a country of Olives and Figtrees and Vines, as well as Corn. The fences, as usual, Cactus and Aloe, but principally the latter. The Palmetta in very great quantities by the side of the sandy road—as common as Furze in England. We ride a little way along the sea-shore ; then arrive at Selinuntium.

The ruins of the three principal temples here, on a step or terrace of land a little way from the sea, are very striking indeed. At a little distance they look very like those great heaps of broken and tumbled rocks which one sees on the tops of some of the Welsh mountains and of the Dartmoor hills. Immense hewn stones and slices of columns tumbled about in wild confusion, scarcely one stone remaining on another, all overthrown and displaced, but scarcely anything broken or missing. Was this caused by an earthquake ? The ruins of the largest temple are perfectly wonderful for the size of the columns and other portions of the architecture ; they are truly works of giants ; the diameter of the fallen portions of the columns (in all these Sicilian temples the whole thickness of the column is in one piece) is more than double my height and said to be in some cases upwards of twelve feet. These columns are not fluted. In some cases the different pieces of them lie close together in due succession, as if they had just slipped down—they give one the idea of giant millstones. The other temples, which are in the same fallen condition, though much 1843 smaller than this, are of majestic size. Their columns are fluted. The stone of which all these are built has the appearance of a shell-sand, or rather, perhaps, coral-sand, agglutinated and hardened. On another tongue of land, perhaps half a mile distant and close to the sea, are other ruins in the same confused state, which *they call* the Theatre; but I could make out nothing of the plan. Here also is a curious hole in the ground (it seems too narrow for a well) lined with a double layer of fine tiles.

All these ruins are intermixed with wild thickets of Palmetta, Mastic, Southernwood, Acanthus, and brambles, which increase the difficulty of making one's way among these heaps of enormous stones. From Selinuntium (where there are no modern buildings but a farmhouse near the great temple and a cottage or two on the other hill) to Sciacea, about twenty-one miles, sometimes along the sandy shore, sometimes over dull, tame clay hills clothed with corn, sometimes over wild wastes covered with Palmetta and Mastic.

Not far from Selinuntium we pass through an open wood of stunted cork-trees worth notice, as the first wild trees I have seen in Sicily. Sciacea, a large walled town, sloping up the side of the hill from the sea, with a castle on the top of a bold cliff above it, has a noble appearance as we approach it along the shore. Like other Sicilian towns, it is best worshipped at respectful distance.

1843   A pleasant sea-breeze made this day's journey much less fatiguing than the last.

Distance called thirty miles.

May 29th. From Sciacea to Monte Alegro, 24 miles, partly along the sea-shore, partly over the long clay hills near it : wild waste tracts which one should be apt to call heaths, but there is no heath on them. The prevailing plants are the Palmetta, the wild Cardoon, a handsome blue Eryngo, a very tall stiff reed-like grass, and a large yellow-flowered plant of the Ferula kind. Not a tree, nor any bush big enough to hide a man. I never saw a more arid country.

Halt in the middle of the day at Monte Alegro ; I never saw a less cheerful place than this miserable dirty village, situated among naked rocks, and scorched by a heat, which I should think could hardly be exceeded in India. The old village, now deserted, was on the peak of a high and exceedingly steep rock ; the present village is at the foot of it. Not far from hence, in the way to Girgenti, enormous rocks of crystalline gypsum, sparkling in the sun like glass. For the greater part of the way to the Port of Girgenti, the hills seem to be mainly composed of this substance. From Monte Alegro to Girgenti is 18 miles. We pass through Siculiana, the filthiest and most miserable town I ever set eyes on,—perched as usual on the top of a hill. The Port of Girgenti has the appearance of a thriving and bustling place. Thence to Girgenti, above three miles, a pretty good road. It was dark when we entered the city, and I could see only that it 1843 stands on a very high hill, and that its streets are narrow.

May 30*th.* The principal Temples of ancient Agrigentum are finely situated on the southern brow of a long ridge, which on that side (towards the sea) descends abruptly in a rough rocky precipice, though of little height. On the other side away from the sea, it slopes gently down into a rich cultivated valley, bounded on the further side by the much higher hill, on which modern Girgenti is built.

The Temple called of Concord, is in fine preservation, and extremely beautiful. Those of Pæstum are more grand, but I do not know that I ever saw a more beautiful building than this. Its columns, fluted, and of more graceful and less massy proportions than those of Pæstum, all remain entire with their entablature and both pediments. The cella is enclosed by walls, which is not the case either at Pæstum or Segesta : these walls built of large (but not enormously large) square blocks of stone, very regular and admirably well joined together without cement. The arches which are cut in these walls, date from the middle ages, when this temple was for a time turned into a church The walls are at the sides of the cella : at the two ends are columns instead of the wall. At each side of either end of the cella, are staircases practised in the thickness of the wall,—repaired, but ancient, leading up to the roof which no longer exists. Resting on the columns at each end of the cella is a sort of (as

1843 it were) inner pediment, and in this a window-like, opening of a very peculiar form, like an isosceles Λ with its top cut off, thus : $\triangle$ But all these details can give no idea of the beauty of the building, which is exquisite.

The Temple of Juno stands at some distance eastward of this, on the highest point of the ridge, in a noble commanding situation. The circeroni here call it the Temple of Juno Lucina: but if it be, as I suppose, the temple mentioned by Pliny, in his story about Zeuxis, he calls it Juno Lacinia. It is in less complete preservation than that of Concord : the columns on the north side are standing, with the architrave and one piece only of the frieze : on the S. side and at both ends, the entablature and the pediments are entirely gone, and the columns want more or less of their entire height.   Here also, the cella was enclosed by walls, the lower part of which remains complete throughout, the upper part has been tumbled down, and lies within the enclosure. Many of the stones are stained of a dark-red colour, which the cicerone attribute to a fire which they say ruined the temple (?)   This temple is raised on a majestic basement of large and regular stones, admirably well put together, and must, when perfect, have been as beautiful as that of Concord, and still more majestic from its situation.   The Temples of Hercules and of Jupiter Olympius are entirely fallen and in ruins.   This last is very remarkable for its enormous size, perhaps exceeding even the great Temple of

Selinuntium. The fallen pieces of columns are of 1843 such magnitude that their flutings are as wide as the shoulders of a moderate sized man, and the triglyphs and mouldings of the cornice astonish one by their gigantic scale. But the overthrow has been still more complete than at Selinuntium : literally, not one stone remains on another. The columns here appear to have been merely half-columns, not detached from the wall. An enormous colossal figure, in an attitude like the *Telamones* in the Tepidarium at Pompeii, lies on its back, unbroken, in the midst of the ruins.

Of the Temple of Castor and Pollux, there remain three columns at one of the angles, with their entablature of remarkable beauty: the rest is fallen. In this temple, and the two I mentioned last, the columns and entablature appear to have been covered with a double coat of white stucco, of which several portions remain ; on the two Temples of Concord and Juno there are no remains of it. The effect of this must have been much inferior in beauty to the fine rich fawn colour of the natural stone.

The temples stand in the following order, reckoning along the ridge from E. to W., or rather from S.E. to N.W. First on the highest point of the ridge, at its eastern extremity, stands Juno ; then Concord ;—Hercules :—Jupiter Olympius ;—Castor and Pollux :—each in succession at a lower level, as the ridge decreases in height. At a little distance, but on another rise of ground, are two columns,

1843 supposed to belong to a Temple of Vulcan; and
below the ridge towards the sea, the tomb of
Theron and the temple of Esculapius. The stone
of which all the temples are built is the same that
forms the hill on which they stand, and also the
much more considerable heights on the opposite
side of the valley. It is a mass of shells and corals,
whole or broken, having a great general resemblance
to the Aldborough crag. I ascended to the high
point which is believed to have been the ancient
citadel. It is on the same side of the valley as the
modern town; it would appear that the ancient city
(which must have been immense), occupied the
slopes of both hills and the valley between.

May 31*st.*   There is a carriage-road, but a very
rough one, from Girgenti to the sulphur mines of
Combatina, above two hours drive from it.

The mines are in the midst of an excessively wild,
bare, and rugged country; they are worked in
the side of a mountain of crystalline gypsum, which
is intermixed with a small quantity of greyish marl.

The sulphur occurs not in veins, but disseminated
as it seems through the whole mass of the gypsum,
hence the mines are not worked by regular shafts
and levels, but are great excavations like caverns,—
at least such was the case with the one I went into,
and I was told the others were similar. The rock of
gypsum and sulphur is cut out in various directions
(seemingly without any regular plan), so as to form
excavations of ample width, and high enough for
a tall man to walk upright in them; thick masses

being left here and there as pillars to support the 1843 roof. When a workman applied his lamp to one of these pillars, I saw the melting sulphur bubble out, and soon a thin blue flame spread over a part of the surface. The mixed substance brought out of the mine is roasted on the spot in kilns, and the sulphur melting runs down into channels prepared for it below, and is cast into large cakes, in which state it is carried down to the port.

The white rocks, the ground stained by sulphurous fumes, the great heaps of raw sulphur and sulphur cakes, the half-naked, brown workmen, and the rugged country around, make altogether a strange and peculiar scene.

The village of Combatina, a very wild-looking place, lies very near to the mines, and at a little distance, on the slope of a hill, the larger village of Aragona.

Not far from hence I find the Sweet Pea growing wild in the margins of the cornfields.

The very curious place called La Macaluba (?) (I am not sure of the spelling) lies a few miles from Aragona, nearer to Girgenti, and between one and two miles off the carriage road. In a wide, bare tract of stiff bluish clay, are a great number of nearly circular holes, full of muddy water, which is in a state of continual agitation from the escape of gas.

In some few cases the mouths of these holes are on a level with the surrounding soil, but in general each hole is in the centre of a little conical hillock, like the cone of a volcano in miniature, rising

1843 scarcely more than a foot above the general level. The largest hole that I saw might be eight or nine feet across; usually they are from a yard down to two or three inches in diameter. In the small ones we very distinctly see the water rise and fall; in the larger the bubbling goes on in two or three points at once, while the rest of the water remains calm.

Sometimes there is a rapid succession of small bubbles as in soda water after the violent effervescence is over; sometimes where the action of the gas is stronger, one sees the mud of the bottom thrown up through the water in great quantities, and every now and then rising above the surface in hemispherical tumours, of perhaps half a foot in diameter, bursting with a very audible noise. The guide told me that the agitation was much greater in wet and stormy weather, and that sometimes the mud was thrown up to the height of 40 palms (about 30 feet.) The water is quite cold, has a salt taste, but as far as I could perceive, only that of common salt, and no perceptible smell. I brought my nose as near as possible to the bubbles when they were most active, but could perceive neither any smell nor the brisk pungent sensation caused by carbonic acid. In some of the holes the water is deeper than I could sound with my walking stick, but in general it is shallow, and at the bottom one feels sometimes stones, more often only mud. I could see nothing in the place which could at all justify one in referring these odd phenomena to volcanic agency, nor could I learn that they had any sympathy with the earthquakes which are felt in other parts of 1843 Sicily.

June 1st. Leaving Girgenti for Palma, we have a fine view of the temples. Descend into the narrow valley between the hill on which they stand, and the hill on which, it is said, the Carthaginians encamped when they beseiged Agrigentum. From thence, a long tract of dreary, bare, waste country, till all at once we descend into the very rich and smiling valley of Palma, in which I noticed the finest Carruba trees I have yet seen. Ascend again to the town of Palma, a wretched place, in the usual Sicilian style. Distance from Girgenti, 16 miles.

June 2nd. The valley of Palma is indeed very pretty and pleasant,—a rich highly-cultivated plain of Olives, Almond and Carruba trees, Vines and Corn inclosed by fertile sloping hills, dotted with trees and crowned with rocks;—altogether quite an oasis in the desert. But after leaving this "happy valley" we have again a long spell of dull, tame, treeless, shadeless, intolerable country, just a ditto to that between Sciacea and Girgenti. Pass through Alicata, a bustling place, which has a considerable trade in sulphur, and which, to judge from appearances, is chiefly inhabited by mules and muleteers. For the last mile or two, travel along a broad but deeps andy road, between hedges of gigantic Prickly Pear. The day's journey, of thirty miles and upwards, ends at Terranova, a town on the sea-shore, seemingly of considerable traffic, but filthy as usual. Here however to my great surprise, I saw a

1843 carriage, and something that really looked like ladies and gentlemen.

June 3rd. From Terranova to Caltagirone, 24 miles : most part of the way through a most wearisome monotonous corn country, without a tree and almost without a flower. At first setting off, indeed, we had a view, though a very distant one, of Etna ; but this was soon lost. Heat suffocating. People cutting the corn in many places ; but it is very odd, with so great an extent of cornfields, one sees hardly any houses. Approaching Caltagirone, we come among much higher hills, and the scenery improves very much. Caltagirone, a large town in a magnificent situation, is far better built and better paved than the generality of Sicilian towns, and has a marked appearance of prosperity. The inn, too, is much less dirty than usual.

June 4th. Caltagirone to Palazzolo, above thirty miles. At first setting out, from the commanding heights on which Caltagirone is situated, had a fine view over variegated hills and wide plains beyond, to the giant Etna, rising in most majestic solitude in the distance. For some miles from Caltagirone, the country is very pleasant, cultivated hills rich in Vines and Olives and all kinds of fruit trees, with hedges of Cactus and Aloe, and variety of pretty flowers by the wayside. Soil sandy, intermixed with layers of a soft crumbly sandstone, containing many shells. This pleasant country continues as far as a miserable village or small town called S. Michele, from whence it changes very much for

the worse ; and after some miles farther, rising 1843
higher and higher among the hills, come to mountains of *lava*, indescribably rough, barren and
fatiguing. I was very much surprised to find lava
at such a distance from Etna ; but the nature of the
rock was unequivocal, and in some places it had all
the appearance even of lava that had *flowed*, the
curdled and concentrically wrinkled appearance that
I had seen in the lava-currents of Vesuvius. For
several miles before we come to Palazzolo, the road
as bad as any I ever saw in my life, not excepting
any in Brasil or the Cape Colony. It seems as if
all the rocks and stones in the island had agreed to
assemble—had given one another a *rendezvous*—in
this road. Once or twice from this lofty region of
stones we caught fine views of Etna and the coast,
the Lake of Lentini, and the Bay of Catania, and
could just distinguish Catania itself. Turning to
the right just short of Buscemi, descend by a path
all but precipitous into a ravine that seems
immeasurably deep, and thence rise up again as
steeply to Palazzolo. In this ravine, however, I
saw a sight that really cheered me, tired as I was—
magnificent trees, Oak and Walnut, not very numerous, but what there were of them really superb.

June 5th. The sepulchral chambers hewn in the
rock, half a mile or so above Palazzolo—the remains
of the ancient Acrœ—are very curious. In the soft,
spongey-looking, tufaceous limestone rock (at least
it appears to be limestone) are a great number
of spacious excavated chambers, each consisting of

1843 several passages crossing one another at right
angles or nearly so, and on the right and left of
every passage are tombs, or, as one should be
tempted to say at first, cribs or bed-places (for they
appear quite suited for such), more than large
enough for a full-grown person : generally two of
these cribs, side by side, between the passage or
ambulacrum and the ranges of pillars of the natural
rock which are left to support the roof, and which
form the divisions of the several parts of the
chamber.

The cribs aforesaid were, when first discovered, each covered by a lid, like the usual lid of
a sarcophagus : those portions of the rock which are
left to form their sides and partition walls may be
from two-and-a-half to three feet high. A considerable quantity of bones, more or less decayed,
remain in the tombs. Every chamber or set of
tombs, according to the antiquaries here, was the
burial place of a particular family; they allowed
room for a good many generations. In one chamber
there are two couple of tombs excavated as it were
in two of the great pillars between the passages, so
that each couple of tombs is insulated, and one can
walk all round it ; moreover, the rock-wall enclosing
it is curiously and elaborately cut out, in a pattern
formed of intersecting of circles.

The mouth of every chamber when first discovered was closed, as I am told, by a great slab of
stone.

The theatre, situated very near to these excavations, is smaller than that of Segesta, but in equally 1843 good preservation, or very nearly so. The ranges of seats, the flights of steps, and the front of the scena are nearly perfect. Adjoining it are remains of a very small theatre, much more shattered, and an exceeding deep well, remarkable as being square instead of cylindrical. The cicerone told me it was 144 cubits deep. The Museum of Palazzolo, I must own, disappointed me ; yet it is a good collection to have been made by one man (the Baron Judica) and in one place, for everything that it contains was found within a very short distance of Palazzolo. This same Baron Judica discovered and opened all the sepulchral chambers, cleared out the well—in short, all that is known about the antiquities of Acræ was discovered by him. In the museum are several fragments of statues; but nothing remarkable in that way; several inscriptions, some painted vases of what the Neapolitan antiquaries call the first epoch, some of the second, and a few of the third; a vast quantity of terra-cotta lamps, lacrymatories and unpainted vases of various shapes; a great number of small heads in terra-cotta, some of them of considerable beauty ; a good many women's hairpins in bronze and silver, some rings, and a few small bronzes.

In the theatre, two or three mills made of lava, exactly of the same shape and structure as those at Pompeii. Palazzolo is a strange place : the rough natural rock serves for the pavement of the streets, and in many cases it forms part of the walls of the

1843 houses, which are built almost as much *in* as *on* it. I note down what have struck me as some of the most obvious characteristics of the south of Sicily in this journey of above 260 miles, from Palermo to Syracuse. The towns (with the exception of Palermo and Syracuse and a very few others) stand on hills, are picturesque and dirty, very good to look at, very bad to go into. Instead of being white, like the Italian towns, they are grey. In the country generally, you see neither hedges nor trees, but either endless unenclosed cornfields or wide waste tracts covered with the dwarf fan-leaved Palm called Palmetto, which is seldom above two feet high.

The general aspect of the country is in a high degree monotonous, naked and parched, though the soil does not appear to be naturally barren. Where there are hedges they are made of the Opuntia (Prickly Pear) or the American Aloe. Single houses are rarely seen : all seem to be collected into towns or large villages. The want of shade, of water, of variety of scenery, of bold natural features and of artificial improvement makes the country very tiresome to travel through. The exceptions to these general remarks I have already specified.

The inhabitants are a very ugly race, more especially the women, and abominably dirty and slovenly in their habits. When Zeuxis inspected all the damsels of Agrigentum, to make up out of their collected charms a picture of perfect beauty, it is to be hoped, for the sake of his picture, that the women of ancient Sicily were handsomer than those of modern. I saw only one tolerably good-looking 1843 woman between Palermo and Syracuse.

The men cover their heads with long white caps, like nightcaps, hanging down on one side—the women generally with black hoods.

Of the natural curiosities, besides the Macaluba, which I have described in detail, nothing surprised me so much as the vast rocks and even mountains of selenite or crystalline gypsum, which occur in many places in the south of the island, particularly near Girgenti. I never elsewhere saw this mineral forming rock-masses. It is not, I suppose, an independent formation, but subordinate to the vast tract of bluish or white marly clay which extends through so great a part of the south of Sicily.

In the animal kingdom, kites and hawks of various kinds are remarkably numerous. The tame swine look like half-starved wild ones. The cattle, dark red or brown, are smaller and much less handsome than the dove-coloured oxen of Italy; but what they want in size and beauty they make up in length of horns. Goats abound and are very handsome. In the west of the island, especially between Palermo and Alcamo, I was much struck with the variety and beauty of the wild flowers, more particularly of the Vetch tribe; but in the arid and monotonous clay country of the south I saw comparatively very few; almost everything seemed parched and burnt up. The most striking botanical features of this clay region are, first, the general prevalence of the Palmetta (Chamœrops humilis), which is as common

1843 here as furze in England; and, second, the great quantity and variety of thistles and plants resembling them. In the whole journey, I saw only two woods—one of stunted cork trees, near Selinuntium; the other of very fine Oaks and Ilexes, a few miles from Palazzolo, on the road to Syracuse. The Oleander occurs in many places on the margins and in the dry beds of streams; in remarkable abundance and beauty in the dried-up torrent-bed along which one travels for a good way, after descending from the mountains, in the way from Palazzolo to Syracuse. It is, however, always a bush, not a tree.

Convolvulus tricolor is common among the corn and on the fallows from Alcamo to Sciacea, or sometimes farther; I saw it sparingly near Girgenti, but not afterwards. A handsome specimen of Hedysarum, with trailing stems and brilliant red flowers, very common in the corn land all through the south. Bupleurum perfoliatum and a smaller specimen of the same genus, likewise very general among the corn. The Sweet Pea I found only near the sulphur-mines of Girgenti.

A beautiful Campion, with bright rose-coloured flowers, abundant near Alcamo, Segesta, and in many places from thence to Sciacea, and again near Caltagirone. I saw very few of the Cistus tribe, no Heaths, no Arbutus, very little Myrtle; Mastic in great quantity among the ruins of Selinuntium and in some other places, but not generally. The Acanthus in remarkable abundance at Selinuntium

and about the temples of Girgenti. Trifolium 1843
stellatum very common in Sicily. Ægilops ovata,
Lagurus ovatus, and Cynosurus echinatus are some
of the most common grasses.

In the sandy country near Caltagirone I saw great
numbers of those curious beetles (the same, I
believe, as the sacred Egyptian Scarabæus) which
employ themselves in rolling balls of dung, in which
they lay their eggs. I did not see them elsewhere

June 7th. Syracuse. The tract of country immediately north of the present town, and which seems
to have been occupied by the ancient city, consists
of very rocky though very low hills, composed of a
soft white tufaceous limestone. The great facility
of cutting this stone led the ancients to make those
immense excavations which give a very peculiar
character to the site of ancient Syracuse. Hardly
anything that remains is *built* or *erected* : all is hewn
in the rock—the theatre, the amphitheatre, the
aqueduct, the tombs, the prisons. The stone
quarries worked by the ancients are immense, and
in some of them have been formed extremely agreeable gardens, especially those of the Capuchins and
of the Marchese Casale, in which the growth of
trees and shrubs is most remarkably luxuriant.
The superb Caruba trees, the Orange and Lemon
trees of extraordinary size, the Olives, Pome-
granates, and Figs, all of the richest foliage and
most brilliant vigour of growth, with the lofty walls
of rock rising above them, and often assuming the
appearance of towers and gigantic buildings, the

1843 huge Prickly Pears crowning the rocks or springing out of their clefts, and the Ivy in many places hanging down from them in festoons, have altogether an effect extremely singular and agreeable. The Carubas and Prickly Pears in these situations, especially in the Marquis Casale's Garden, are the largest I have seen anywhere.

Some of the large Fig-trees, growing through the clefts of the rocks or spreading over their surface, have a most singular appearance. The Papyrus grows in great quantity in the Cyane (Fonte Ciane); but I understand there is none of it in the Anapus, either above or below the junction—certainly I did not see any. The guides say, very probably with truth, that this is owing to the difference in the quality of the water. The Cyane is a strong and clear stream, though very weedy. The Papyrus grows in very large and very dense tufts, as much as ten to twelve feet high, in some places forming entire islands. The marsh vegetation of the banks of both these streams is exceedingly dense and luxuriant : Cyperus Longus and Polypogon Monsp. grow here to a .larger size than I have seen anywhere else.

I saw an extraordinary quantity of Dragon-flies, principally of the kind with blackish wings and a glossy dark green or bronzed body.

The Theatre of Syracuse is much the largest ancient theatre I have seen, but not in such perfect preservation as those of Segesta and Acræ : the seats (all cut in the soft calcareous rock) are unfortunately interrupted by a water-mill, which has been 1843 constructed in such a situation as to cut the *cavea* right in two. The general plan, however, is visible enough, and the situation, as usual with the ancient theatres, such as to command an extensive and striking view.

Fountain of Arethusa. Situated within the walls, near the extremity of the island occupied by the modern city. It is a good copious stream, supplying a large tank, which is generally full of ugly women busily employed in washing, with their clothes tucked up remarkably high. As soon as a stranger appears on the terrace overlooking this tank, the women with one accord begin to beg; and if one throws down any small money among them the scrambling and chattering are very droll: I never saw the like except in the Zoological Gardens. The water is not muddy, as Mrs. S. says, but clear enough when free from the disturbance occasioned by the washing.

The fluted Doric columns which are built into the wall of the Cathedral (supposed to have belonged to the Temple of Minerva) are immense.

The Museum inconsiderable : the chief thing in it is a Venus in white marble, very beautiful, though unluckily wanting the head and one arm. It is in much the same attitude as the Venus of the Capitol, but has a little drapery, which she is holding up with the remaining hand, but which, however, hides very little. The forms are very fine and the execution delicate and highly finished.

1843   [I can find no further journal nor any more letters in 1843, only a statement that he returned to England in August.   He went to London in February, 1844, and the 27th is the first date on which he makes an entry in his journal.—F.J.B.]

1844.

—

1844   *February 27th.*   Dr. Falconer tells me that in the Himalaya Mountains, which he has explored from Nepaul to near Cabool, nothing is more remarkable than the mixture of different types of vegetation characteristic of various and remote regions.   He found representatives of the Floras of Europe, of Siberia, of Arabia and Egypt, and even of South America.

The vegetation of Affghaunistaun is more like that of Persia than of India, and has something of the Desert character.   In those spots of the Indian Desert where there is a little vegetation, the variety of forms collected in a small space is very striking. In one such spot, he tells me, Mr. Edgeworth collected 24 species, which belonged to not less than 19 different natural orders.

Dr. F. was struck with the great general resemblance between the vegetation of the Indian Desert and that of Egypt.

Dr. F. found Pyrola Rotundifolia and P. Secunda 1844 in the Himalayas.

I went with the Lyells and Miss Horners and Dr. Falconer to the British Museum to see the fossil remains which have been lately added to that collection. There are a great number of remarkable extinct animals from the Sub-Himalayan Hills: in particular, Dr. F. showed us the skull and teeth of a species of elephant, which he considers as exactly intermediate between the elephants and mastodons, and as proving that those two supposed genera are really the same.

Also bones and portions of the shell of a land-tortoise of most gigantic size. It was interesting to compare these with corresponding parts of a recent tortoise, which he showed us side by side with them, and which agreed exactly in everything but size. There is also, from the same hills, a perfect shell of a fossil tortoise, which seems to be the very same species with one actually existing in India. The gavial, or crocodile of the Ganges, identical with the existing species, is found in the same deposit, among many extinct forms. Another crocodile (an extinct species), which seems to have been 40 feet long.

Two kinds of horse, one of which resembled the Arab in the beautiful form of its head.

Many of these organic remains were imbedded in brown iron-stone, so that much labour and care were required to extricate them.

The fine skeleton of the mastodon, which was

1844 exhibited two years ago at the Egyptian Hall, under the title of the " Missouri Leviathan," has been purchased for the Museum, and set up according to the true rules of anatomy, so that it looks much more natural than before.

Lyell tells me that Berzelius, who analysed specimens of nearly all the *meteoric* stones known, found in some one or other of them (besides the iron and nickel which are the chief ingredients of them all) minute portions of all the known metals and metalloids : in one, for instance, a minute portion of gold, in another of silver, in another of copper, and so on.

Dr. Falconer says that there are two if not three different kinds of crocodile in the Ganges (besides the gavial, one of which he has seen above 30 feet long).

March 6th. British Museum. The Xanthian marbles, brought from Lycia by Mr. Fellows, are the most important of the recent additions : they are curious rather than beautiful. It is evident that they are of two or three different periods. The basreliefs from the " Harpy Tomb" (figured in Mr. F.'s book) are evidently of a very early style, and supposed indeed to date from as far as a hundred years before Phidias. They are in very low relief, and strongly marked with the stiff, hard and formal characteristics of early art.

One is struck with their general resemblance in style to the Persepolitan sculptures, of which there are casts in the same room. In some particulars also they remind one of Egyptian art, particularly

in the long narrow eyes, which in the profile are 1844
represented as if seen in front. All the faces are
entirely in profile ; the mouths (as in the Egina
marbles) turned up towards the corners ; the
drapery in very straight and formal folds. Still
these sculptures are much less rude and grotesque
than those of the Selinuntium metopes.

Other friezes, from the same place but from different buildings, are very clearly of a later date and
an improved style of art. They represent combats
of warriors on foot and on horseback, and the
assault of a town : some of the combatants wear
Persian caps, others helmets with immense crests ;
but all are clad in long flowing garments almost
feminine. Lastly, there are portions of some beautiful female figures in flowing tunics, unfortunately
much mutilated, in which there remains very little
trace of the *archaic* stiffness.

Among the many beauties of the Townley Gallery, my especial favourite is the nymph resting
after the chase—half lying down and resting upon
her left hand. Nothing can be more sweet and
graceful.

March 14*th.* Went with the Lyells and two Miss
Horners to see Sir William Symonds' model room
at Somerset House. It is very interesting. Besides
a vast collection of models of ships, and of everything connected with ship building, there is a
special collection which interested me particularly—
specimens of every kind of wood that could be procured, especially those used in shipbuilding, together

1844 with dried specimens of the leaves, flowers and fruit
of the tree to which each belongs. A great many
from New Zealand in particular.   Among these the
most remarkable is the Kowdie or Kowrie tree.
Dammara Australis—an immense tree singularly
valuable for masts and spars.   Sir William told us
that one of the ships of his building has a top-mast
of this wood, which has been in use 13 years, and it
is as sound as ever ; whereas the ordinary duration
of a top-mast of common pine timber is not more
than three or four years.   The Dammara has cones
much like those of the Cedar of Lebanon, and
oblong coriaceous leaves a good deal resembling
those of some species of Leucospermum.   Its resin,
which becomes very hard, with much of the appearance of pale coloured amber, and is very fragrant
when burnt, is found not merely in large masses, but
in very extensive layers, beneath the surface of the
soil in the parts of New Zealand where this noble
tree abounds.

There are here abundant specimens of the Mahogany tree, which has very large woody seed
vessels, and of the Teak, with its fine large leaves
and wide-spreading panicles of flowers ; fine specimens of the Cedar of Lebanon, brought from
Lebanon itself by Sir William Symonds's son ; and
a great variety of the Pine and Oak tribes.   Of all
the varieties of Oak timber, Sir William told me,
the most valuable is that of the Tuscan Oak ; from
the specimens of it, gathered by Sir William himself
in the upper valleys of the Apennines, it appeared

to me to be a variety of Quercus sessiliflora, with 1844 more deeply cut leaves. It has certainly nothing to do with Quercus Cerris, which, as Sir William told us, is quite useless for ship-building. The Tuscan Oak has not mossy cups, nor are the lobes of its leaves pointed. The wood of the Douglas Fir is good for nothing, its grain being remarkably open, porous and spongey.

Among the timber trees of New Zealand, of which specimens are here preserved, one is a Fuchsia, Fuchsia excorticata; its wood is beautifully satiny.

Speaking of the ancient galleys, Sir William Symonds gave it as his decided opinion that the common notion of their having several successive ranges or *banks* of oars was erroneous; that the management of the upper ranges of oars in such situations would be impossible. He believes that the terms " bireme," " trireme," " quinquareme," and so forth implied merely the number of men working at each oar.

The first ship built in England on scientific principles was the Royal Sovereign in the reign of Charles the First.

March 22nd. Went into St. Paul's, where I had not been (inside the church I mean) since 1829. Much struck with the gloomy, cold, poor and naked effect of the interior as compared with that of St. Peter's, which is fresh in my recollection. It is true the day was dull and cloudy; but I have often been

1844 in St. Peter's in equally gloomy weather, which in no degree destroyed its gorgeous beauty.

The Puritanical dislike to painting and rich decoration in churches, which has too long prevailed in this country, has been a grievous enemy to art and taste. Yet I conceive that a very strict religionist might think monuments to warriors and commemorations of bloody battles as ill-suited to a Christian Church as painting and gilding.

The monuments in St. Paul's are more interesting for their subjects than admirable in design. The most to my taste are those which have merely a statue of the great man commemorated, without any allegorical figures. I saw none which could bear a comparison with Canova's monument of Clement the Thirteenth or Thorwaldsen's of Pius the Seventh. Even Bernini's fanciful tomb of Alexander the Seventh is better than the insipid Fames, Victorys and Britannias which prevail in St. Paul's.

March 28th. Dr. Lindley thinks that, among the British Roses, there are four forms which may be considered as satisfactorily distinct species, viz. :— R. Spinosissima, Rubiginosa, Canina and Arvensis. Perhaps R. Villosa (or Mollis) may be considered as another. All the rest, he is satisfied, are varieties of these four or five, or in many cases mules between them. The Cape Orchidaceæ, he says, are uncultivable : several of them have at different times been introduced into this country, and have flowered once ; but no one has succeeded in keeping them above one season. The only exception is the great

Satyrium Corifolium, which has sometimes been found to thrive tolerably well. The Lowea, or Rosa Berberifolia, is cultivated with tolerable success in the Horticultural Society's Garden, and flowers not very sparingly; but no way has been found to propagate it.

[On the 10th of April, a beautiful day, he made an expedition to Kew with the Charles Lyells and myself. I remember his saying that existence alone on such a beautiful day in these gardens was happiness. He writes in his journal on the 11th:—" The day of my supreme happiness." This was the day we were engaged, and there is no further entry in his journal for some weeks.—F.J.B.]

1844

LETTERS.

From his Mother's Sister, Caroline Lady Napier, Wife of General Sir William Napier.

Havilland Hall, Guernsey,
April 20th, 1844.

My dearest Charles,

The news contained in Emily's* last letter made me so happy about you that I cannot keep silent till I hear from yourself, but must tell you, my very dear nephew, how truly and warmly and affectionately I rejoice in your happy prospects and sympathize with you. I have deep cause to feel for you and to love you, for your beloved mother's sake and for your *own*, as well when you were an engaging delicate child looking up often to your aunt for help and amusement as since you have

* Lady Bunbury.

1844 grown up to be the kind nephew and the sensible,
amiable, estimable being you now are. God bless
you my own dear Charles, and grant you all the
happiness I wish you and think you deserve. I like
extremely all I hear of my future niece, and shall
rejoice when the time arrives to be introduced to
her. And meanwhile I want to hear many more
particulars about her and about all your plans and
expectations, and when the marriage is likely to
take place. Write me a long account and do not
fear to tire me : every particular will be interesting
on the subject. I think she bears a Christian name,*
one of the dearest to me that the world contains.
Your uncle and the three cousins who are at home
join me most cordially in affectionate congratula-
tions. Perhaps you will see Louy in her way
through London, and Catty also.

Catty is going to dearest Bessie, who has been
very unwell and is very weak and delicate since
she got back to Ireland. As there is no post
from here till Monday, I will not close this till then.
I heard two days ago from Henry,† telling me he
expects to be in England in May, somewhere
between the 16th and 20th. You will have heard
all about us from Emily, and your time and
thoughts must both be full now.

I am ever, dear Charles,

Your truly affectionate Aunt and Friend,

CAROLINE NAPIER.

---

* Name of her eldest daughter, who died some years before.

† Charles Bunbury's brother.

Your letter, though dated the 15th, only reached me to-day, dear Charles : it is a delightful letter and just like yourself. I have not the *least* doubt you will make a good husband, and that in all that depends upon you, your dear Frances will be *very* happy. I do long to know her. Could you not take the Channel Islands in your wedding tour ? I do not feel as if my letter was half as warm as I feel.

From his Father.

Barton,
May 28th, 1844.

My Dear Charles,

This is my last letter to you as a single man; you are now going to set up in business for yourself; and I hope with full sincerity of heart that you and your partner may be prosperous through life. It is but due to you, in quitting your apprenticeship, that I should bear my cordial testimony to your " past service."

You have been a very affectionate son, kind and considerate towards me at all times; nor have you ever given me pain by your conduct or stirred my anger further than a passing vexation.

I hope that Frances and yourself will enjoy your tour thoroughly, and that you may do so the better I pray that this vile weather may melt into a real June at the instant when the clergyman shall join your hands.

Ever most affectionately yours,

H. E. BUNBURY.

From Lady Bunbury.

My Dear Charles,

Your father has allowed me to read his
letter to you, and most heartily do I concur in every
kind wish and every word of praise he has written.
I love you most sincerely for your own sake, but
above all for the strong filial respect and affection
I have ever seen in you towards him whom I love
best on earth.

Your affectionate Mother-in-law,

E.B.

———

From Sir Henry Bunbury to Miss Horner.

My Dear Frances,

You and I are coming into very near rela-
tionship.   As yet we are not acquainted ; but I hope
that before long you will come to know me and may
like me better than you have any reason to do as
yet.   I trust and I believe that you will make my
son happy ; and by so doing you will command my
gratitude and establish the best claim to my affection.   I am rather an odd old gentleman, and am
not accustomed to make professions ; but you will
find me sincere and as straightforward as you can
reasonably expect from a descendant of Adam.
I hope you will like me, and I am very well disposed
to like you : so if you please we will try.

In the meantime, my blessing be upon your
marriage.

I beg to be kindly remembered to Mr. and Mrs. 1844
Horner and all your sisters, and you are to believe
me from henceforth,

Very affectionately yours,

H. E. BUNBURY.

Barton,
May 28th, 1844.

I am going to Mildenhall this week to get the old
house into a decent state to receive you on your
return from the North.

———

May 29th, 1844.

My Dear Father,

I am just finishing my arrangements for
departure, and have not time to write you a long
letter; but I will not neglect to thank you, which I
do most heartily, for your very kind letter which I
received this morning. I deeply feel your kindness
and am truly happy that you have been satisfied
with my conduct. To you, indeed—to the excellent
education which I have received from you and my
dear mother and to your constant attention to my
welfare and improvement— I owe more than I can
express, and I trust you will never find me insensible of it. To Emily also I owe a great deal—her
affection and kindness have been unremitting, her
advice always excellent; and if I have not time to
write to her separately on the present occasion I
hope she will not suppose me ungrateful for her

1844 kindness, which I feel very much. I am satisfied that I have as good a prospect of happiness in marriage as any human being can possibly have, and I trust that the faults of my character may be to a considerable degree corrected by the responsible situation in which I shall be placed—feeling as I do that I have charge of the happiness of one who is most dear to me, who relies upon me for guidance, support and protection. I am very sorry to hear that Emily has again been ill since her return to Barton. Pray give my best love to her, with my warm thanks for her note.

Ever your very affectionate Son,

C. J. F. BUNBURY.

P.S.—Our direction will be till about the 10th of June, " Post Office, Keswick" — afterwards, " at Charles Lyell's, Esq , Kinnordy, Kerriemuir, Forfarshire."

Chesham Place,
June 12th, 1844.

My Dear Charles,

Your letter and especially its happiness gave me great pleasure. God bless you both, and may it long continue unbroken by sorrow or sickness, for from no other cause do I apprehend any interruption to it, suited as you seem to be in tastes and opinions, and united by strong mature affection. I think I have always read your character right,

dear Charles, and I have at last shown I was not 1844
mistaken in believing it—one formed to feel and
appreciate domestic happiness in no common
degree!

You will be glad to hear that my brother* has
returned in perfect health and looking *one day
younger* than when he left us seven years ago! He
and Lady Napier desire their kindest congratula-
tions to you and Frances, and were delighted to
hear of your marriage. My Cissy is now sufficiently
recovered for me to take her out of town the day
after to-morrow, and right glad I shall be to get out
of the heat and bustle of this town, which, when one
has neither spirits nor time to enter into its
pleasures, is about the most odious place upon earth
in my opinion. However, I have had the pleasure of
seeing Lady Campbell† and dear Miss Fox‡ in my
transit, the latter *very* well and Pamela as delightful
as ever. Tell Fanny (for that is to be my name for
her) that I saw Mrs. Horner and all her sisters
yesterday, looking very well, though some were in
the agonies of *packing* and preparing for their
departure to-day.

I *was* very sorry not to be at your wedding for
better reasons than missing the *big-wigs*, though
I should have liked to meet them too.

Your's affectionately,
EMILY BUNBURY.

---

* Sir George Napier, on his return from the Government of the Cape.
† Pamela Lady Campbell, the daughter of Lord Edward Fitzgerald.
‡ Caroline, sister of the third Lord Holland.

1844   [We were married on the 30th of May.   I here
add a letter from Lady Lyell to Mrs. Ticknor, with
a little account of the wedding, forwarded to my
husband some years afterwards by my sister, Mrs.
Lyell.—F.J.B.] :—

My Dear Charles,

In copying out one of Charles Lyell's letters to
Mr. Ticknor, I have come on a P.S. from Mary to
Mrs. Ticknor, which I think you will like to hear,
though it is an *old story !*—"Kinnordy, 15th, June,
" 1844.— I have to thank you for two kind letters ;
" your last I have sent to my sister *Mrs. Bunbury* (who
" is now on her wedding tour), to show her the kind
" interest you take in her.   She was married on the
" 30th of May, and a prettier wedding I cannot
" imagine.   She was dressed in white watered silk,
" with a wreath of orange flowers sent by my aunt
" from Paris, and a lace veil which I wore on the
" same happy occasion.   She looked very lovely
" and was wonderfully composed.   My four younger
" sisters were bridesmaids, all in white muslin over
" silk, with a white rose in their hair.   The marriage
" took place about eleven, and a great many friends
" met us in the church, and at twelve we sat down
" to breakfast in my father's house—about 40.   The
" bride and bridegroom at the top of a horse-shoe
" table, and the rest of the company placed round.
" Among those you know were Mr. and Miss
" Rogers, Mr. and Miss Hallam, Mr. Babbage, Mr.
" Sedgwick, and there were a good many relations

"of both sides. My husband proposed 'The 1844 " ' Health of Bride and Bridegroom,' and Mr. " Bunbury returned thanks. I thought both speeches "just what they ought to be ; but there was too " much feeling in both not to overcome us very " much. Mr. Sedgwick made an excellent speech, " and restored our equanimity a good deal. He had " known Mr. Bunbury at Cambridge, and he is a " great favourite of his ; so there was much feeling " there also. At two o'clock my sister changed her " dress, and set off in a britzka for a tour to Oxford, " Warwick and the Lakes, where I presume they " now are. In a week we expect them here to pass " a fortnight. Frances has been a great deal here, " and is much attached to all my husband's family, " and it will be a great pleasure to us to see them."

––––––

[Sir Henry and Lady Bunbury were unable to be present at the wedding owing to engagements at Barton, and Cecilia Napier, though in London, was in too delicate health to be able to come ; but Edward Bunbury came, and our dear friends Mr. and Mrs. Richard Napier came, with Augusta, daughter of Captain Henry Napier (afterwards Mrs. Frederick Freeman) ; Mr. and Mrs. Nicholson and their two daughters (afterwards Lady Galton and Mrs. Bonham Carter) ; Florence Nightingale and her sister Parthenope (afterwards Lady Verney) ; my uncle and aunt, Mr and Mrs. Lloyd ; my dear governess, Miss Parker ; Mr. and Mrs. Frederick Maurice ; Mr. Robert Brown, well known as the

1844 Prince of Botanists ; Lady Bell, the widow of Sir Charles ; my cousins, Mr. and Mrs. Pellew ; Mr. Charles Mallet, grandson of Mallet du Pan ; and William Loch.   After our happy tour to the Lakes it was delightful to meet at Kinnordy the Charles Lyells and my dearest sister Susan ; then in Edinburgh we saw a good deal of many old friends, especially my legal ones—Lords Cockburn, Jeffrey, Rutherford and Cunningham ; then we went to Leeds to see my husband's very dear cousin, Sarah Clarke, and there we met her father and stepmother, Sir George and Lady Napier, with whom we travelled to town.   Here we spent some happy days with my dearest father, mother and sisters, and here I was first introduced to Charles' brother Henry, afterwards Col. Bunbury.   We went down to Barton on the 6th of August, and spent a happy time there before we settled at Mildenhall.   Here we again met Sir George and Lady Napier.

LETTER.

My Dear Father,

My No. 7 was written on the 7th of last
month, just before I set out on my return from
Graham's Town; I arrived here quite safe and in
high health on the 28th of June, and I can assure
you I was most happy to find myself at home
again.

We (that is Captain Dundas and I) were very
fortunate in our journey, having fine weather almost
the whole time and being delayed only one day, yet
both of us were heartily weary of the journey
before it was half over. My general impression of
all I have seen of this colony (beyond the first 30 or
40 miles from Cape Town, I mean) is, that is a
country which I am glad to *have seen* once, but
which I never should feel the slightest inclination to
re-visit. Never before did I travel so far through a
country so generally barren, monotonous and
dreary; there are a few redeeming points, but they
are very few. No wonder the colony is poor and
thinly inhabited: I do not see how it can ever be
otherwise, till some way can be discovered of
cultivating the earth without water, or of making
bread out of stones. Even where the soil is good
for something, there is such a want of capital and
skill and enterprise, and the communications are

d

1838. consequently so wretched, that the produce cannot be turned to good account.

I found the two girls looking gloriously well on my arrival here, and as you may suppose, they gave me a warm welcome. Jack, too, seems to be quite recovered and in very good spirits, quite a different creature from what he was six months ago : he is a noble fellow, exceedingly clever, ambitious and as generous, high-minded, and *chivalrous* (in the best sense of chivalry) as any body I know. I cannot help feeling sorry that he should be in a profession which holds out so deplorably unpromising a prospect—so little chance of distinction, as the army does in these times.

Craig is gone up to Graham's Town to join George Napier, who I am afraid is likely to be detained an indefinite time at that vile place.

Mr. Martin West, of whom I believe you know something, has been appointed Civil Commissioner of Albany, and what is stranger, seems delighted with the appointment, which I am sure is more than I should be; I would as soon undertake the management of a hornet's nest. It happened very oddly: he had actually sailed for England, but the vessel was so much damaged in a gale of wind that she was obliged to put back into Simon's Bay, and Mr. West, after running the most imminent risk of drowning, found this appointment waiting for him when he went on shore. It is said he is well off in point of money, and only desired employment of which he will now have enough to satisfy anybody.

I have seen lately a good deal of the Elliots, and 1838. spent some days very pleasantly with them. They are very kind and hospitable. Miss Elliot is beautiful, and her manners the most winning that can be.—I think her a very good and amiable girl : and Sarah, who is a far better judge in such cases than I can pretend to be, thinks so too.

The Admiral seems to be a good and sensible man, but I do not yet feel well acquainted with him.

I met with an odd accident when I was there the other day : we all went to a ball on board the Scout, and in coming away, going down the ship's side in a hurry and almost in the dark, I missed the boat, lost my hold of the hand-ropes, and fell plump into the water, but I was picked up directly, and very fortunately have caught no cold. I have met with several acquaintances of Hanmer's on board the Malville and Scout.

By the Cruiser, I have a satisfactory letter from *young* Emily, who seems to have got really well again, but from no one else.

By the last English papers which have arrived (up to the 12th May), I see that the Ministers were *still* in, but that a violent attack was preparing against them on the ground of the Appropriation clause, which, if successful, would, I suppose, compel their resignation.

Best love to Emily and to my brothers.

Ever your very affectionate son,

C. J. F. B.

LETTER.

Government House,
October 3rd, 1838.

My Dear Father,

I thank you very heartily for your kind
letters of May 8th and June 30th, and for the
supplies of newspapers which you sent me.   It is, I
find, upwards of two months since I last wrote to,
you, but I have in that time, written two letters to
Emily, which where in fact intended as much for
your information as for hers, and which of course
you will have seen.   At this present writing, I
have no news to send you from this quarter of
the world, so that I shall be reduced to comment
upon what I have read and heard from England.—
I have been much struck with an article in the
Edinburgh Review (by Lord Brougham, I understand) on George the Fourth and Queen Caroline,
and on the abuses of the press.   The historical
part of it is very interesting, the characters are
drawn with a life-like spirit and vigour, and with a
delicacy of discrimination, which seem to me to
rival Clarendon's best portraits, and what pleases
me most of all is the way in which he points out
the tendency of a royal education, and the influences
which made the Prince what he was.   As to the
remarks on the prevailing licentiousness of the press
in attacking private character, I am very much
inclined to concur in them, rather than in the
counter arguments ot the examiner.   Brougham,

1838.

1838. after his long eclipse, seems to have blazed out
again, both in speech and in writing, as brilliant,
and as eccentric as ever.   His speeches on the
slavery question are most eloquent ; but I can by
no means approve either of his interference in
behalf of the Glasgow cotton spinners, or of the
unholy alliance which he seems to have formed on
many questions with Lord Lyndhurst.   I confess,
notwithstanding the good he has done in some
cases, he is not a politician whom I like, or in whom
I feel any confidence; there is something in his
conduct which looks to me so much like faction, and
as if vanity were his his ruling motive.

As you say, it is very strange how the ministers
continue to hold their places ; it seems almost as if
they were protected by their very weakness; the
Tories perhaps see that things are not ripe for their
own accession to power (at least for their secure
possession of it) and therefore may think it their
best policy to leave that power, for the present, in
the hands of a set of men who are ready enough, in
all conscience, to make concessions to them, and
whose measures they can control and modify by
their opposition.   Certainly, if I had been in
Parliament, I must have voted against the Ministry
on several questions.

It would seem however that those long-fought
questions of the Irish Tithes and Irish Corporations
are likely to be settled at last, by a compromise.   I
speak according to the latest accounts I have seen
from England, which came down only to the end of
June.

We are all well here : at present the two 1838.
girls and I are alone, John being absent on an
excursion to the Paarl, about 40 miles distant. It
will not be many days, I hope, before the General
and the rest of the party arrive, but we do not
exactly know where they are just now, as none of
them wrote by the last post.

I have been a good deal with the Elliots lately,
and like them more and more as I know them
better ; I am in hopes they will be our neighbours
in the country during the hot season, when Simon's
Town, as well as Cape Town, is deserted by all who
can get away from it.

I like the Admiral extremely, now that I have got
really acquainted with him.

I must not forget to thank you for your
book, of which I have read one part (the Memoir)
with a great deal of interest and satisfaction ; most
of the letters I had seen before at Barton, and I
must confess I think that some of them were hardly
worth publishing, though there are many which are
are interesting.

I hope you will not give up your plan respecting
the Italian military adventures.

There have been races here, to which I went one
day (the first time I have been at a race since I was
a child), but a duller affair I seldom witnessed.

Pray give my love to Emily and my brothers,
and believe me,

Ever your very affectionate son.

C. J. F. B.

P.S. The General will be here on the 10th of this

1838. month.   I have given up the scheme of going to
Bombay ; but I do not purpose to leave this before
March, as that is in all points of view the best
season for returning to England.

LETTER.

Government House, Cape Town,
December 6th, 1838.

My Dear Father,

You will no doubt have heard before this, 1838. of poor Sarah's illness, which, though not alarming, has been tedious and distressing, and has much damped the cheerfulness of our family party. She is now, I hope, really recovering, though still weak, but there seems reason to fear that she will be liable to a recurrence of similar attacks every hot season. It is very fortunate that she has a high opinion of, and perfect confidence in the medical man who attends her (Dr. Clarke of the 72nd), who indeed is in every respect worthy of confidence and esteem. All the rest of the party are very well.

John Napier and Major Charters sailed some time ago with a company of the 72nd, and some artillery and engineers, to take possession of Port Natal, and they are probably arrived there before this, but we have not as yet heard anything from them.

Charters is a great loss to us in the way of society, being one of the most agreeable men I know,—an enthusiast in science, full of information and acuteness, and strikingly bold and original in his views, though often (I think) hasty and unsound. I must tell you Col. Bell's pun on the subject of this expedition: he said that the Natal settlers wanted a constitution, but that the Governor,

a

1838. instead of sending them a *new constitution*, sent them *old Charters*.

I hope to hear from you soon: the last letter I had from you was of June the 30th; I have heard, indeed, rather more recently from Mr. Eagle, but his letter (of July 16th) gave me no information about you, and, indeed, very little information on any subject. I am very anxious to know what you think of the real strength and effect of these *demonstrations* in favour of Universal Suffrage, which make such a figure in *The Spectator:* if that paper is to be believed, they are very formidable, but they may be more so in appearance than in reality. I hope my friends at Bury will not take up this cry: if they make Universal Suffrage an essential article in their political catechism, they must look out for another candidate, for I decidedly will not vote for swamping the votes of the middle and upper classes and giving an overwhelming preponderance to the most needy and least educated portion of the community, and those most swayed by passion and by temporary impulses.

The first of this month was the day on which all the apprentices in this colony became free: it passed off without the least disturbance, and every thing is going on so quietly that a person who did not *know* that the change had taken place, would perceive no outward symptoms of it. I understand that the apprentices in general shew the best dispositions, and are very ready to work; many of them, of course, quit the service of their former masters for other employments in which they can earn better wages, but the number of idlers seems to be rather diminished than increased, and I am told that the number of committals in this last week has been less than usual. The colony altogether seems to be in a tranquil and comparatively prosperous condition: the depredations of the Caffers have undoubtedly diminished and the value of property has risen.

I have been reading the Duke of Wellington's "Dispatches" with a great deal of interest. It is a most important and remarkable work (though I think the want of arrangement very inconvenient), and gives to me at least a higher idea of the Duke's abilities and greatness of mind than I ever had before; it is true that the substance of the information contained in it is the same with that I had before read in my uncle's history: but in following his proceedings and the train of his thoughts thus in minute detail, and from day to day, one is more impressed with the strength and clearness of his mind, his great and comprehensive views, his firmness amidst such difficulties and embarassments as I really believe no other man could have successfully struggled against, and the unselfish spirit with which he was ready to sacrifice even his own fame to what he believed to be the good of his country. By the way, I must explain myself: when I say the *want of* arrangement, I mean the mixing up trifling or merely technical matters, — remarks on Courts-martial, orders to commissaries, points of military etiquette, &c., with the documents of real importance.

1838.   I have been very sorry to hear that Dr. Somerville
is considered in danger : I should regret his loss,
not only on his own account, but because I fear
that even in a pecuniary view it would be a heavy
blow to his family, and would break up what I
always found one of the pleasantest houses in
London.

I wrote to Henry last month, I hope he will not
have rejoined his regiment before I return to
England.

Pray give my love to Emily and my brothers, and
believe me,

Your very affectionate son,

C. J. F. B.

LETTER.

Government House, Cape Town,
December 29th, 1838.

My Dear Edward,

I wish you a merry Christmas and a happy new year—though the latter is not quite begun yet. I suppose you are now availing yourself of the Christmas holidays to get out of town, but in whatever part of the country you may be, I guess you are not complaining either of heat or mosquitos, as we are here; during the greater part of the month indeed we have had uncommonly cool, even cold weather, which has contributed much to Sarah's re-establishment, but now the hot weather which was to be expected at this season of the year, has set in with a vengeance. We have heard from Charters and John Napier from Port Natal—all well up to the 13th of this month, in high spirits and delighted with the place, although exposed to excessive heat and swarms of insects. They had landed without opposition or accident, and were busy at work constructing a fortified camp. The Boors encamped in that neighbourhood had received them civilly, but are said to be in great distress, their crops having failed, and their horses and cattle dying fast of sickness, so it would seem that Natal is no such fine country after all.

Another portion of the emigrant Boors had marched on commands against Dingaan, the king of the Zooloos, and there are many conflicting specu-

b

1838. lations as to the result of this expedition, but
nothing is yet known with certainty.   The arrival of
the letters from Charters and Jack was a great
satisfaction to us all, as some Dutch rascals in the
eastern part of the Colony, with their usual malig-
nity, had got up a story of a fight between the
emigrants and the soldiers, and of the total
destruction of the latter ;   and though few people
believed this, yet it made the General very uneasy.

You will of course have heard all about Sarah's
illness.   I am happy to say she is now well again,
though not yet quite strong enough to sing.   We
have been for these two months, expecting to move
into a house in the country, about five miles from
hence, and I believe it is now nearly ready at last.
The air is much fresher and healthier there than in
town, and it is said there are no mosquitos, *here*
they are almost as troublesome as at Rio, and there
is likewise good store of fleas in Government house,
a shocking fact, but too true.

In the way of reading, I have been chiefly
employed of late with the Duke of Wellington's
" Dispatches," a work of great interest and importance in a political as well as a military view, and
one which has given me a still higher opinion of that
great man than I had before.   I have got through
four volumes and a half of it, but it is a tough job
and will last me, I dare say, all the time I remain
here.   Before I began this I had been much
entertained with Abercrombie's book on the "Intellectual Powers," particularly the very ample and
interesting discussions on the curious subjects of

dreaming, somnambulism, and insanity. It is 1838. about the clearest and most attractive metaphysical work that I have met with, though not the most profound. I have also re-read Southey's "Progress and Prospects of Society," (a favourite book of mine) and read a good deal in a less connected way. I do not think I shall meddle with Arnold's " Rome," until I have done with the Duke.

I hope your tour to Scotland and the Hebrides pleased you, and that you collected lots of minerals, and curious rock specimens.

By the way I never saw a country so utterly uninteresting in that respect as South Africa.

And how goes on the collection of coins ? I suppose you have added largely to it since I left England. In truth I shall be very glad to get home again ; though the kindness of my friends in this house has prevented me from ever feeling that painful sickening, longing for home, which tormented me so much in Brazil, and though I delight in their society, I am quite tired of the Cape itself, and should but for them have hated it long ago. I do not know any country so much over-rated. I shall stay on however till March, so as to arrive in England at a tolerable season of the year. I mean to return home direct, and not go round by Bombay.

Believe me ever,

Your very affectionate brother,

CHARLES J. F. BUNBURY.

APPENDIX.

PALACES OF ROME.

I have in another place written down in detail my 1844 observations on the picture-galleries I have seen at Rome. I shall here only recapitulate them very briefly.

Borghese. A very large and very fine collection of pictures, conveniently distributed through a great number of handsome rooms. Those that chiefly struck me are :—

1. A magnificent portrait of " Cæsar Borgia," by *Raphael.*

2. " St. Dominic," by *Titian.*

3. A warrior in armour, with a young man behind him, called " Saul and David," by *Giorgione*— a picture of extraordinary effect.

4. " Deposition from the Cross"—*Raphael :* the female figures admirable, but some of the men are in forced and awkward attitudes, with an unpleasant appearance of effort.

5. The same subject, by *Benvenuto Garofalo :* a very fine composition, free from the usual stiffness of *Garofalo,* but gloomy in colouring.

6. " Sacred and Profane Love"—*Titian.*

7. " Madonna and Child," by *Carlo Dolci :* an excellent specimen of his peculiar style.

8. " Diana and Her Nymphs"—*Domenichino.*

1844   9. " Danæ," by *Correggio.*

10. " Leda," of the school of *Lionardo da Vinci.*

Sciarra Palace. 1. " Modesty and Vanity," by *Lionardo da Vinci* : one of the most charming pictures I have ever seen.

2. A very fine portrait, by *Raphael*, commonly known as" Il Suonatore."

3. " The Gamesters," by *Caravaggio :* excellent.

4. Portrait of a woman, by *Titian*, known as " La Bella di Tiziano."

5. Two pictures (nearly repetitions) of " The Magdalen"—*Guido* : very good specimens of his *pale* style.

6. Some capital little pictures by *Breughel.*

Barberini Palace. 1. The famous portrait by *Raphael*, called " La Fornarina," but entirely unlike the one so called at Florence : an ugly and vulgar woman, with a very bad countenance, but most beautifully painted.

2. " Beatrice Cenci"—*Guido.*

3. A fine portrait of a Lady, by *Titian.*

4. " Joseph and Potiphar's Wife," by *Biliberti :* full of expression.

5. " Madonna and Child"—*Francesco Francia.*

6. The same subject, by *Innocenzo da Imola* : very Raphaelesque.

Rospigliosi Palace. *Guido's* lovely " Aurora," a fresco most worthy of its reputation, is on the ceiling of a summer-house or pavilion in the garden of this palace ; and in the same pavilion are a few oil-paintings worthy of mention :—

1. "Sampson Pulling Down the Temple," by 1843 *Ludovico Caracci* : a grand picture, though disfigured by some incongruities.

2. " Adam and Eve in Paradise," by *Domenichino* : a most singular and even whimsical composition.

3. " The Triumph of David after his Victory over Goliah"—*Domenichino*.

4. Heads of the Twelve Apostles, in as many separate pictures, by *Rubens*. There is also a fine antique bust of Scipio Africanus the elder.

Farnese Palace. In this great palace, the principal and almost the only objects of attention are the frescos of *Annibale Caracci* on the ceiling and end walls of one of the rooms, and very beautiful they are. Their subjects are various amatory stories from the ancient mythology, such as " Venus and Anchises," " Hercules and Omphale," " Acis and Galatea," &c. The two larger compartments at the ends—" Andromeda Chained to the Rock" and " Perseus Turning His Assailants Into-Stone"—are to my taste less satisfactory than the rest.

In the court-yard of this palace is a sarcophagus, which is reported to have been found in the tomb of Cecilia Metella ; but the tradition is doubtful, and there is nothing on the sarcophagus itself to identify it.

Farnesina Palace. The ceiling of the entrance-hall is covered with the famous frescos of " The Story of Psyche," designed by *Raphael* and painted by some of his scholars, with portions here and there finished by his own hand. They are in excel-

1844 lent preservation, and some of the groups are very fine, especially the first two of the series and that representing the appeal of Cupid to Jupiter. In a good many, however, the female figures appear to me coarse, too muscular, and the colouring too red: this is more particularly the case in the two larger compositions in the middle, which are by *Giulio Romano*, a very disagreeable painter to my taste. In an adjoining hall are *Raphael's* "Triumph of Galatea" (which disappointed me much), and a magnificent colossal head sketched in charcoal by *Michael Angelo.* In an upstairs room, two beautiful frescos by *Razzi*, of Siena—"Alexander Entering the Tent of Darius" and " The Marriage of Alexander with Roxana."

Colonna Palace. Here I was much more struck with the extraordinary splendour of the gallery itself than with any of the pictures. There are, nevertheless, many fine portraits and several other pictures which almost anywhere else would attract much attention; but at Rome one grows fastidious from the profusion and variety of excellence.

The chief things I noticed here were two curious little pictures by *John Van Eyck*, the inventor of oil-painting; two very clever caricatures of Italian peasants at dinner, the one by *Annibale Caracci*, the other by *Caravaggio ;* " The Magdalen Carried to Heaven by Angels," by *A. Caracci ;* and a most grotesque old picture, by a certain *Niccolo Alunno*, of Foligno, wherein the Virgin is most vigorously belabouring the Devil with a big stick.

Doria Palace. An immense collection, bewildering 1844 by the multitude of pictures, of very various degrees of merit, and a great part of them in bad lights. I can name but a few, though I daresay there are many others deserving of praise :—

1. A portrait by *Lionardo da Vinci*, called "Queen Joanna of Naples," but it can hardly be either of the celebrated Queens of that name ; however, it is certainly one of the most lovely portraits I ever saw.

2. A masterly head of a Monk, by *Rubens* : it is said to be his Confessor.

3. Two portraits, " Bartolo and Baldo," in one picture, by *Raphael.* Who were Bartolo and Baldo?

4. " A Holy Family," by *Sassoferrato*—charming.

5. " The Meeting of the Virgin Mary and Saint Elizabeth"—*Beni. Garofalo.*

6. Some magnificent *Claudes*, perhaps the finest I have ever seen.

7. " Belisarius," by *Salvator Rosa* : a grand specimen of the master.

8. " Misers Counting Money," by " The Blacksmith of Antwerp" *(Quintin Matsys).*

9. Five *Breughels*—" The Terrestrial Paradise" and "The Four Elements": admirable in their way.

There is a fine specimen of this master, *Velvet Breughels*, at Barton Hall,—" Pilgrims at a Shrine."

FINIS.

Milton Keynes UK
Ingram Content Group UK Ltd.
UKHW041521181024
449640UK00009B/105

9 781108 041126